Essays on Lighting

Essays on Lighting

J. G. HOLMES
ARCS, BSc, DIC, CEng,
FIEE, FInstP, FSGT

*Honorary Member of the
Illuminating Engineering Society,
Formerly Technical Director, Holophane Ltd.*

CRANE, RUSSAK & Company, Inc.
NEW YORK

First Published August 1975

© J. G. Holmes 1975

All rights reserved. No part of this publication may be reproduced, stored in a retrieval system, or transmitted, in any form or by any means, electronic, mechanical, photocopying, recording or otherwise, without the prior permission of Adam Hilger Limited.

Published in the United States by
Crane, Russak and Company, Inc.
52 Vanderbilt Avenue
New York, N. Y. 10017

Library of Congress Catalog Card No. 75-21733
ISBN 0 8448 0771 0

Published in Great Britain by
Adam Hilger Ltd.,
Rank Precision Industries
29 King Street
London WC2E 8JH

Made and printed in Great Britain by
The Garden City Press Limited
Letchworth, Hertfordshire SG6 1JS

*Dedicated to my wife
who asked me to
keep it simple*

Contents

Preface   xi

1 Light   1

2 Lighting   7
The right place for the lamps – The right direction for the light – The luminous environment – The quantity of light

3 Measuring Light   16
What do we measure? – Direction as well as amount – Flux and luminance – Units, abbreviations and terms

4 Dirt and Maintenance   27
The effect of dirt – The cost of dirt – The cost of lamp replacement – Maintenance – The cost of maintenance

5 Colour — 36
Coloured light – Coloured things – Colour appearance and colour rendering – Defective colour vision – Description and specification – Measurement of colour

6 Light Control — 46
Control of light by reflectors – Control of light by refractors – Control of light by diffusers – Control of light by absorption – Optical materials

7 Safety and Durability — 55
Dangers in use – International specifications – British specifications – Protection against electric shock – Marking – Electrical and mechanical design – Heat dissipation – Standards for particular fittings

8 Lamps — 64
Sources of energy – Output of energy – Incandescence and tungsten lamps – Effects of voltage change – Light-controlling lamps

9 Discharge Lamps — 73
Electric discharge – Luminescence and fluorescence – Temperature effects – Effects of voltage change – Lamp performance

10 Scalar and Vector — 82
Illuminance and illumination – Scalar illuminance – Illumination vector – Planar illuminance – Cylindrical illuminance and conical illuminance – Why do we need these ideas? – Flow of daylight – Flow of electric light – Vector/scalar ratio

11 Daylight Design 92
Calculation of amount of daylight – Measurements and predictions – Interpretation of daylight factors

12 What we want from Daylight 101
The problem of quality – Windowless buildings – Analysis of quality – Daylight glare – Natural and artificial lighting

13 Hours of Sunshine 110
The sun's movements – The sundial – Sunpath diagrams – Probability of sunshine – Building design – Further reading

14 Road Lighting I 118
How it works – White, blue or yellow light – Light distribution – Wet roads – New techniques

15 Road Lighting II 126
How much light? – Bridges and humps – Junctions and roundabouts – Pedestrian crossings and white lines – Underpasses and tunnels

16 Foggy Days and Misty Nights 134
How much light? – How much contrast? – Yellow light in fog?

17 Motorway Fog 143
How we see when moving – Driving in fog – Motorway fog

18 Stage Lighting 150
How much light? – Seeing the actors and seeing the scene – Lamps and equipment – Controlling the light – Planning the lighting – Colour – The open stage

19 Plastics 159
Names – Constitution – Properties – Effects of heat – Effects of light and ultraviolet radiation – Electrical properties – Conclusion

20 Glass 167
Composition and properties – Mechanical strength – Optical properties – Thermal properties – Permanence

Envoi 175

Preface

Light is essential to seeing and is as common as the air we breathe, but it has several very complex technologies. The processes of seeing are so natural and familiar that they are rarely understood. This book attempts to examine some of the basic concepts of natural light and electric light and to express them in simple terms.

The twenty essays were published over the period January 1972 to April 1974 in the journal *Light and Lighting*, under the general heading 'Let's keep it simple'. They were directed towards those readers who were concerned with using light for seeing but whose primary interest was outside the lighting profession.

As with other professions, lighting has its own jargon and rules of thumb which are accepted and used by the professionals, sometimes without much thought for the fundamentals on which they are based. As the essays developed, the needs of students and of architects became merged with those of some lighting professionals who may use light as a commodity without proper respect for its unique qualities.

The author is indebted to many colleagues who have borne with his persistent philosophizing and particularly to Gordon Chamberlin, Eric Harper, Ken Jackson, Ray Jennings, Francis Reid and Walter Stevens who contributed directly to the content of some of the essays. Thanks are also due to The Illuminating Engineering Society for permission to publish the essays in this form and for the use of their Code and Technical Reports to which frequent references have been made.

J. G. Holmes
December, 1974

1 Light

Light is power, streaming as radiation from the sun or a lamp, and bouncing off all the things around us until some of it reaches our eyes. This enables us to "see" the things around us.

The conversion of other forms of energy such as electricity into this particular kind of radiation is what we call the "production of light" and is the principal concern of the lamp technologist. The process of bouncing the light in proper proportion off the things around us comprises all the arts of "lighting design", from modelling to colour rendering. The reaction of our eyes and our brains, which gives us sight and comprehension, comes within the study of "vision".

In passing, we notice that the scope of the Illuminating Engineering Society covers all these studies, light, lighting and vision. Each of these three has to be linked to the other two if it is to have any real meaning. The IES studies many other things as well, such as heat and psychology, but these do not concern us at this moment.

Light is radiant power. It suffers only two kinds of change between its source and our eyes; it may be re-directed many times (from one straight line to another) or it may be absorbed. We shall see later that both these changes have to occur before we can see the world around us. The re-direction we call "diffusion" and it can occur either by reflection or transmission. The absorption may be partial (as at a grey surface) or total (at a black surface) but the light continues until this happens; after absorption the power in the radiation is converted into heat. Thus lighting is a dynamic process, with energy continually flowing from its source and filling the whole visual environment until it suffers its death by absorption, to be resurrected as heat which also has to flow somewhere by conduction or convection. There is another form of resurrection, by fluorescence, the conversion of short wave light (blue or ultraviolet) into longer wavelengths (red-orange or white light), but this is really the production of new light.

Fortunately the speed of light is so great (300 000 kilometres per second) that all this seems to happen instantaneously, and the wavelengths of light are so short (about 500 millionths of a millimetre) that they may be neglected in comparison with the sizes of everyday things. We are not conscious of the flow of light as a matter of velocity or time, neither are we conscious of the wavelength or frequency of the radiations of which it is composed.

We are taught that light travels in straight lines. This is true for the ideal case of a hypothetical "ray" of light or the exceptional case of a laser beam, but it is not the whole truth. Diffused light, which is what we have in everyday life, may be considered as a bundle of rays, each travelling in a straight line but whose *average* direction may change so that light may seem to flow— and in fact does flow—around corners. As an example, light from the sky entering a window has a downward direction but in the middle of the room the direction of the light is horizontal, away from the window and towards the back wall, whilst in the upper parts of the room the flow of light is upwards, on to the ceiling. The principal path of the light from window to ceiling is curved, as may be simply shown by examining the shadows of a small object placed in different parts of the room. This may be difficult to understand but it is due to the partial absorption of the more strongly downward rays which fall on the floor, and to the partial reflection from floor and furniture, etc. within the room which has a strong upward component; the flow of light is the average flow of all its component rays. It is a good thing that this happens because it is basic in the achievement of a proper distribution of illumination within an interior.

Light is defined as "radiation capable of stimulating the eye"; it exists whether or not we are there to see it, but it is assessed in terms of its ability to enable us to see. Only a narrow band of wavelengths has this property, approximately one octave in the middle of the whole electromagnetic spectrum extending from X-rays to ultraviolet, visible and infrared and into radio waves, a gamut of wavelengths from about 10^{-10} metre to 10^4 metres, all emitted to some extent from the natural source of radiation, the sun. The wavelengths of light extend from violet light of 380 nanometres to red light of 760 nanometres (a nanometre is a convenient unit equal to one-millionth of a millimetre). It is interesting that this narrow range of wavelengths is at the peak of the power distribution emitted by the sun and that the atmosphere, which transmits light very freely, is partly opaque to the ultraviolet and the infrared beyond the limits of the visible part of the spectrum.

The amount of power involved in light radiation is very small. One lumen of white light is about 0·004 of a watt and one lumen of the yellow light from a sodium lamp, because it is near the peak of the sensitivity of the eye, is about 0·002 W. A practical sodium lamp has a conversion efficiency of about 25 per cent; that is to say, 25 per cent of its electrical power consumption is turned into visible light. The maximum theoretical conversion efficiency for a whitish light is about 40 per cent and a practical mercury halide lamp producing white light gives about 15 per cent which is comparable with the performance of a sodium lamp. So even though the amounts of power are small, the efficiencies of modern lamps are remarkably high.

We can read this page very easily by the light represented by 0·05 W reflected from the whole page, requiring under optimum conditions only 0·15 W of electrical power. Most of the light reflected from this page goes into the rest of the room and not into our eyes; the amount of power actually entering our two eyes whilst reading this page may be about 10^{-5} W. It is on this tiny stimulus that our eyes and brain work. Putting it another way, light is so important to us that our eyes have been made almost incredibly sensitive, able to detect delicate differences of colour, shape or texture with a total stimulus of the order of a microwatt.

Apart from looking directly at a lamp, light itself is invisible in the sense that a ray of light passing through empty space cannot be seen. Light has to be reflected or diffused by some object so that it—the object—can be seen; the sun's rays through the window or the beam from a searchlight are invisible unless they are revealed by the dust and droplets in the air.

The particular value of light to us is that it can be changed, in colour or intensity, by the objects which reflect (or diffuse) the light falling on them; this is how we see these objects. A piece of ice under water does not appreciably change the light passing through it and is almost invisible. As a generalization, we see things because they absorb some of the light and reflect or diffuse the rest. A cloud or a snowflake are exceptions to this generalization, because they absorb scarcely any light, but to carry the example further, it is the very absence of absorption in mist or snow that makes us virtually blind if we experience a "white-out" in a snow storm. Unless the light around us creates contrast, it is of no value to our sight.

There are two principal modes of differential absorption—if the power in some of the wavelengths in the spectrum is changed, by partial absorption, we see a change in colour; if the intensity of the light is changed, by partial absorption or by a directional effect, we see a difference in contrast. The skill of the lighting engineer lies in providing the conditions in which the light can react with the things around us in both these ways so as to reveal them to best advantage.

We usually think of light as "white", probably because most natural or artificial light sources produce white light. But white is a compromise colour. It can only occur when the light contains a reasonably well-balanced distribution of power in the spectrum of the visible wavelengths, 380–760 nm. If the balance is towards the long wavelengths, as in light produced by a flame, we say it has a warm colour; if towards the short wavelengths, as in the light scattered by tiny particles in the upper atmosphere, the colour is cold; these terms indicate our psychological associations with red-rich and blue-rich light. Light is in fact coloured, always, and the component colours can be analysed very precisely in terms of wavelengths, which is the basis of the science of spectroscopy and colorimetry. White or pink or purple lights are compromise colours because they do not occur amongst the pure colours of the spectrum but are produced in the eye by the combined response to a range of different wavelengths. This is a subject in itself and for the moment we need only note that light can only produce colours which are already contained in it—the selective absorption by a "coloured" object cannot *add* a colour to the light. Thus the pure yellow light of a low pressure sodium lamp contains no other colour and it makes both pillar boxes and policemen look yellow or brown. The same is true of the pure red of a No 6 theatre spotlight colour, but most coloured lights are not pure in this sense; they are not what we call mono-chromatic and they contain a mixture of pure colours. Fortunately, the white lights from lamps or the sun contain mixtures of all the pure colours. Let us note in passing that a white surface such as this paper is a different kind of whiteness from white light; white paper reflects all wavelengths equally and consistently within a few per cent, whereas white light includes nearly all visible wavelengths but not equally or consistently. This is why fluorescent tubes and the like can be designed to give us such a variety of "white" lights.

We usually think of light as diffused and multidirectional, whether from the sun and sky or from luminaires and ceiling and walls. At any position in space, there is an amount of light (the total luminous flux reaching that point) and there is a flow of light (the complex of directions from which the light reaches the point). It is obvious that the effectiveness of the light in enabling us to see depends on both amount and flow. It is not enough to measure the amount of light irrespective of its directional effects; in fact, the directional effects are probably more important than the amount. When this light falls on an object, the amount is reduced by absorption, slightly for a whitish object or considerably for a dark object, and this can give us contrast on a scale from about 80 per cent to 2 per cent. The flow of this light is changed by reflection, refraction or diffusion, which can give us contrast in a much more delicate way and on a much wider scale. The capacity of light to reveal texture and shape is just as important as to reveal colour or shade.

To summarize, this mysterious element we call light is a flow of radiant power which is virtually imperceptible unless it is changed in spectral balance (colour) or in amount (absorption) or in direction (diffusion and texture) and which is invariably absorbed almost instantaneously by the objects on which it falls. When we see an object, it is because the light falling on that object is changed in direction or amount or colour by the object. Light is our means of visual perception of the whole world around us but we only use for vision less than one ten-millionth part of the power from our lamps, or even less of natural light. Yet this is the substance of the lighting industry, it is one of the basic elements of the natural world and it provides one of the best means of communication and information known to us. This is why the phrase "reverence for light" is a key thought in the science and technology of illumination.

2 Lighting

There is more to lighting than the lumens on the working plane, which is merely a convenient figment of the illuminating engineer's imagination, an approximate measure of the quantity of useful light but no guide to the suitability of the lighting. Quality is no less important than quantity.

Lighting is the application of light to an object to enable us to see; it is the process of illumination, the design and the control of the distribution of light and of the visual effects created by it. The sole object of lighting is to enable us to see. Full account must therefore be taken of the personal, subjective reactions of the people for whom the light is provided. There are commercial incentives in this because people do not work properly or play properly unless they can see properly—most of our daily work is done with the aid of artificial lighting and the balance of economic advantage depends on suitability no less than on amount.

It would be discourteous to the lighting industry to compress the whole technology of illuminating engineering into a few pages, so this chapter does not attempt to be a contents list for a textbook on lighting. It omits the production of light, the effects of colour and the arithmetic of calculation although all these are important.

The principal objectives of lighting design are to put the lamps in the right places, to control the flow of light in the right directions and to create the right luminous environment. The mechanics of lighting which serve these objectives are the quantity and design of the luminaires, the calculation or measurement of the illuminance and, for the built environment, the colours and reflectances of the surrounding surfaces. The results are achieved by a combination of instinct and experience which has been investigated intensively this century but which will defy detailed scientific analysis for many years yet.

The right place for the lamps

The choice of the "right" place for the light sources is determined by convention, by the surrounding structure and of course by the object which is to be lit. We often arrange this almost automatically for ourselves because we can move the object or move ourselves to suit the lighting—we "take it to the light". It is more difficult to choose the right places for other people, particularly if we do not know the detail of their visual task. The light which falls nicely over one person's shoulder may be all right for him but wrong for his neighbour; the lamp in the middle of a room illuminates all the walls (and into the cupboards), but for many people it throws the shadows in the wrong places. The lighting for a particular object can only be suited to everybody if all are facing the same way and looking at the object, as in a theatre or football match; in general the only way of lighting a multi-purpose area such as an office or a railway station is from above, so that the conditions are equally well suited—or ill suited—to everyone. This means a principally downward flow of light, which has the merit of simplicity, but in enclosed spaces this is accompanied by many multiple reflections and shadows. It is usually worth while to arrange things so that for each person the dominant lamp is near the best place for his particular task.

The right direction for the light

The directions in which light is emitted from each lamp are controlled by the optical design of the luminaire: straight down in a narrow beam from a BZ1 "downlight"; widely spread from a BZ6 "general diffusing" luminaire; up and down without much sideways spill from a direct-indirect luminaire; and mostly upwards from an indirect luminaire. The BZ numbers are the classification introduced by the Illuminating Engineering Society for the calculation of utilization and of glare index; they are not figures of merit but indications of the breadth of the spread of light, so that the larger the BZ number, the higher the proportion of light which goes directly on to walls and vertical surfaces. BZ1 for localized lighting, BZ2 or BZ3 for a low glare environment, BZ5 or BZ6 for a bright and lively environment, BZ8 or BZ9 for economical lighting out of doors or at low levels; these brief recommendations are an over-simplification because each area or each object of regard ought always to be considered on its own merits. It has been suggested that BZ2 or BZ3 is a good general-purpose distribution, principally downwards without too much sideways light, but this experience is based on the current fashion of high reflectance surfaces.

Lighting is not all direct from lamps, except out of doors where there may be no walls or roof to reflect the light. As a rough guide, about one-quarter of the light at any particular place comes from diffuse reflections at the walls and ceiling. This reflected light is of paramount importance in softening shadows and in illuminating vertical surfaces or under surfaces of objects. It is this reflected light which can flow round corners in ways which direct light cannot. In a large area, reflected light may be reinforced by oblique light from more distant sources but this may be associated with a degree of glare.

Although the principal flow of light in most interiors is downward, it is the cross flow component which largely determines quality and which can greatly enhance the visual pleasure of a scene. An environment filled with direct light only may be harsh and unsocial, even though the lighting may be efficient and free from glare. Cross lighting only is likely to be pleasant but bright, glaring and probably inefficient. Some compromise is necessary to ensure a proper balance between the downward and sideways components. An example of this effect is a room lit through rooflights only, which give strongly-directed downward lighting, or through windows only, which give strong cross lighting; a combination of these can yield satisfactory lighting over a large area. PSALI (permanent supplementary artificial lighting of interiors) is another example of this technique.

In social environments the use of wall brackets, table lamps or semi-indirect luminaires promotes the cross flow of light. It seems natural to do this, particularly when the illuminance is relatively low. The good habit of lighting a domestic area from lamps near the corners of the room has the same effect. The present day fashion for ceiling-mounted lights giving a high illuminance on horizontal surfaces has been made possible by the cheapness of light but it is at the expense of the aesthetics of the visual scene. Lighting engineers have not yet found simple methods of calculation or description or even of appraisal to enable the engineer's consistency to improve on the artist's experienced intuition. There are no firm rules, and certainly no equivalent to the "lumen method" of calculation, for predicting the balanced cross flow of light in an interior.

The flow of outdoor lighting is quite different. The absence of reflecting surfaces makes it necessary for the cross flow to be provided by direct light and we find that vehicle headlamps, street lanterns and sports floodlights all shine horizontally or obliquely instead of directly down. This again is natural—sunshine is usually oblique and we find that natural scenes look most pleasant in the morning or evening when the sun is low. It must not be overdone: unidirectional lighting or horizontal lighting can cause disturbing effects. An observer looking down the beam of a vehicle headlamp or a searchlight may find that things are brightly lit but the lighting is featureless, the object looks flat and the shadows are confusing. For an observer looking across the beam or towards the lamp, the lighting is much less effective in terms of brightness and the shadows may be so dark that there is none of the gradation which usually conveys the idea of shape—the lighting engineer would say of both conditions that the modelling is poor. Good outdoor lighting requires separated light sources with overlapping beams and therefore with graded shadows, to give solidarity and realism to the scene.

This leads us to another generalization, that the flow of light on to an object should be such that we can recognize it readily and can see its details sufficiently, without conscious mental effort and without the necessity of learning a new technique of vision, if the lighting is to achieve its purpose of enabling us to see. As an example, the obstructions on a traffic road may actually be revealed in stronger contrast at night by ordinary street lighting than in daylight, but there is no doubt that we can recognize them more confidently, and negotiate them more safely, in the daytime when they have their "normal" appearance. This generalization is equally true indoors and out.

The luminous environment

"To enable us to see" requires more than the illumination of the object of regard; the whole luminous environment should be in balance. Fortunately our eyes are remarkably tolerant. Excessive contrasts or the wrong distribution of light and dark areas or excessive brightnesses can however impair our vision or, more often, cause real tiredness and complaint.

Our surroundings contribute to our vision in two ways which are related to the two kinds of thing we want to see, namely, a particular object of regard or the general disposition of everything around us. Firstly, there is the reflected light which is an important part of the lighting of an object and there is the general background which determines the level of sensitivity and the state of tension in the automatic muscles of our eyes. The reflected light affects the modelling as already discussed; the adaptation level affects our scale of brightness—whether the object seems too brightly lit or whether the shadows or dark parts of the object are too dark. The eye muscles are the only parts of the eye which can become tired—this occurs if they are overworked, either consciously or unconsciously. Secondly, the whole surroundings must be lit in the right proportion to reveal their shape and to maintain our frame of reference, to reveal their character so that we know what is going on, and to provide interest beyond the content of the particular object of regard. This is what decides whether the environment looks bright or dull and whether our attention is concentrated on the object or wanders around the field of view. It follows, obviously, that the designer of the lighting ought to know what the observer wants to see before he starts to plan a lighting installation; this very rarely occurs and most lighting installations are of a somewhat general-purpose type. There is a rough rule of thumb that the luminances of the object, of its immediate background and of the general surroundings should be in the ratio 100 : 30 : 10 (this is discussed in an Appendix of the IES Code). In practice, conditions vary so widely that no such simplification can be applied and this "rule" is rarely observed in lighting design. It is more relevant to reflectances than to luminances and it may sometimes give useful guidance in the choice of surface finishes. Our eyes are phototropic—to borrow from biology a word which means "seeking after light"—and we tend involuntarily to look at the brightest areas in the field of view.

Glare is an unwanted feature in any lighting installation other than a fairground. Disability glare impairs our ability to see but may be avoided by ensuring that there are no unduly bright areas near the objects of regard. Reflected glare also impairs our ability to see, either by the reflection of bright light sources near the directions of regard or, much more often, by surface reflections in the very thing we are looking at, so reducing the contrasts by which we see. It is interesting that we need contrast to see but it is excessive contrast that causes glare. Discomfort glare may be tiring or disturbing but it does not impair vision if a deliberate effort is made to disregard it; the IES Glare Index is a nominal figure assessing discomfort glare. In practical situations, we are immediately conscious of disability or discomfort glare if we bother to give it a thought but it is not easy for us to detect reflected glare. This merits more thought than is usually given to it because the decreased visual performance associated with poor contrast rendering can be much more significant than that due to a substantial reduction in illuminance. Putting it the other way round, the poor contrast rendering caused by reflected glare requires a disproportionate increase in lighting level to compensate for it and, if this is provided, the reflected glare may itself be increased for some other directions of view. It is often better to switch off the light which is in the wrong place and which causes reflected glare, even though this reduces the amount of light available for seeing. To summarize, disability glare is rare in modern lighting installations, discomfort glare is controlled by observance of the IES Glare Index limits and reflected glare is something to be concerned about in any particular location.

The quantity of light

The amount of light, the illuminance level or the luminance of the object, is a principal criterion in assessing whether a lighting installation is "sufficient and suitable". It can be measured or calculated with fair precision, it can be related to visual performance and it can be costed. With all these quantitative facilities, it has become a yardstick of "good lighting"; if we assume that lighting is suitable in quality, then the amount determines the sufficiency in quantity. This is not always so. If the quality of the lighting is not suitable in terms of wrong distribution or glare as already discussed, then the more there is, the worse is the result. One might live and work in a small office with a single bare 60 W lamp but the visual conditions would become impossible with a 500 W lamp, because the quality was wrong for either lamp. Most modern artificial lighting installations are designed, or have evolved, to work well at the illuminance level at which they operate. It would often be desirable to alter the design, particularly in terms of light distribution, if the level were to be greatly changed, up or down. When Ward Harrison asked the question in 1937: "What is wrong with our 50 foot-candle installations?" he found that the greatly increased illuminance levels in the USA at that time had been produced without any change in the technique of lighting; the same kind of mistake still happens when glare and the flow of light are not properly controlled.

The amount of light required for working situations, as recommended in the IES Code, is determined partly by the degree of severity of the visual task (i.e. the luminance and contrast levels to achieve a certain performance level) and partly by the current fashion. Fashion is dependent on many factors including economics, amenity, staff turnover, prestige and so on, each of which is subject to statistical trends which always seem to rise rather than fall. Study of the IES Codes since 1936 shows an average rise of 4 per cent per annum, which is less than the average rise in the efficacy of white light lamps. Social prediction techniques do not indicate any reduction in this general rate of growth and many modern office buildings have installations well above the IES figures, but there is a physical limit to the increase in efficacy of lamps; although the IES recommendations have not in the past called for much in the way of increased power consumption, this state of stability will be disturbed by the pressure of social trends. It follows that the amount of light for any particular visual task depends on the requirements of the task itself in relation to an amenity or performance scale which in 1972 was still rising at about 4 per cent each year.

Lighting techniques have kept pace with the developments in lamp technology and in social awareness of the value of light, but the safety margins between good lighting and bad lighting are narrowing as light has become cheaper and more plentiful. The correct location of lamps, the optimum distribution of the light emitted by the luminaires and the design of the whole visual field become more critical as the amount of light is increased.

3 Measuring Light

Our ability to see depends largely on the amount of light available, so we like to know how much we have. Particularly, electric light costs money which is always measured with great precision, and we need to know how much we are getting for our money. It is interesting that daylight, which is nominally free, is usually measured in relative terms (daylight factor, daylight variation, etc.) so that few photometers are ever calibrated to read daylight flux or illuminance directly.

We cannot measure the effect of lighting any more than the effect of a coal fire could be expressed in terms of the calorific value of the fuel or the number of hundredweights burned. What we measure is the physical amount of light but we need to take care because a "light meter" makes it seem so simple and precise that we may be deceived into thinking it tells the whole story.

Measurement of light has developed to a higher level of accuracy than in many of the other features of our environment. The run-of-the-mill light meter will measure correctly to about ± 10 per cent which may not seem particularly good to an accountant but is very precise in comparison with the enormous range of lighting levels (over 10 000 to 1) we can easily tolerate. Consider what happens with our other senses. A noise level meter should be right to ± 2 decibels which is roughly ± 40 per cent in sound power or about a 2·5:1 ratio. Temperature seems to be measured very accurately with a thermometer calibrated in degrees but the heat transfer from a radiator or the heat balance at a point are only measured to ± 10 per cent (or better under laboratory conditions); an accuracy of $\pm 1°C$ in ambient air temperature is not particularly close in comparison with the range from say $-10°C$ to $50°C$ which we can easily tolerate. More important in economic terms, the dietician is hard put to it to measure our daily food intake in calories to better than ± 10 per cent although the range between starvation and excess is less than 1:10. And who can measure taste or smell to ± 10 per cent? Even though the purist may say that photometry is an inexact science, the photometrist with his ultimate absolute error of perhaps ± 1 per cent need not be apologetic to anybody.

What do we measure?

A photocell, the usual device for photometry, measures the total light flux which falls on it. As its area is constant it can be calibrated in flux per unit area, which is illuminance, in lux.

For the reader unfamiliar with these terms, there are two words which need to be distinguished: *illumination* is the process of lighting an object so that it may be seen and *illuminance* is the amount of light falling on the object in flux per unit area. If the distance from a single light source is known, the measurement can become illuminance $\times$ (distance)^2 which is the intensity of the source in candelas. If the collection angle of the light falling on the photocell is restricted and known, the measurement can become illuminance/(solid angle) which is the luminance of the area "observed" by the photocell in candelas per square metre or in apostilbs.

The common light meter is designed to measure illuminance and is calibrated in lux. If a meter is calibrated in lumens per square foot (foot candles) the reading should be multiplied by 11 to convert it to the value in lumens per square metre (lux).

Visual photometers may be used for luminance measurements, usually by an equality-of-brightness match with an adjustable brightness of an internal lamp. This is convenient because the eye responds to luminance rather than to the total amount of light or to illuminance, and because a reliable match can be achieved over a wide range of luminances and of near-white colours.

It is important for us to know what we are measuring because the photocell cannot tell us anything more than the amount of electrical power created inside it. Firstly, the colour of the illumination must be known because the cell responds differently from the human eye and a correction factor for each particular illuminant is needed if reasonable accuracy is to be attained; this factor may range between $0 \cdot 8$ and $1 \cdot 4$ for good quality cells and present-day lamps. Secondly, the direction of the light falling on the surface of the cell should be considered, because most cells respond less than they should to light rays incident at more than $60°$ from the normal; we must remember that more than half of the visible environment is in the zone between the horizontal and $60°$ from the vertical. Thirdly, shadows from obstructions must be considered; the ordinary observer is completely opaque and almost non-reflecting so that he may cause a significant reduction in the reading of a light meter just by being close enough to read it. Stray light, from unwanted reflections or from other sources which ought not to be included in the measurement, should also be considered and eliminated. Fourthly, the supply voltage, the ballast rating, the ambient temperature and the amount of dust or dirt may have significant effects on the result. Fifthly, but not least important, we should know when the light meter was last calibrated and with what result. It is not unusual for a light meter to give readings which are 20 per cent or so less than the correct reading, so regular checks on the sensitivity and calibration are very necessary if the results are to have real meaning.

Direction as well as amount

Rays of light have direction as well as amount and in many situations the direction of the flow of light is no less important than the amounts of flux or illuminance involved. We can, of course, estimate the directions fairly accurately from experience and we do not need directional photometers for general-purpose measurements. But we must take account of the direction of flow of light in order to get a meaningful result for the amount of light.

A horizontal photocell on the working plane in an office is well suited to the principally downward flow of light from the usual type of installation but it gives little information about the lighting of the walls and other near vertical surfaces forming most of the field of view. Where the lighting is designed to be oblique, as in floodlighting of any outdoor area or in some industrial situations, the measurement should take into account both the direction of the light and the orientation of the surfaces receiving it; this often becomes so complicated that the problem is solved either by common sense, on the basis that what is near enough will do, or by detailed measurement and calculation. Street lighting is an interesting combination of these techniques; very few measurements are required to ensure that an installation complies with the Code of Practice but the most elaborate measurements are involved in the approval of a lantern or the calculation of road surface luminance.

The angular distribution of intensity from a luminaire is important in determining its usefulness in terms of utilization of the light and avoidance of glare; this is the basis of the BZ system. The intensity distribution from a symmetrical luminaire (note that most good recessed luminaires with prismatic "cut-off" panels give a symmetrical distribution even if the panel is long and narrow) can be represented by a vertical polar curve; this is a diagram in which a radial distance from the centre to the curve represents both the intensity of the light in candelas and the direction in which that intensity is emitted. Such a measurement requires an apparatus in which the photocell is swung in an arc around the luminaire, either physically or by means of mirrors, and the luminaire is also capable of being turned on its own axis. This will enable the intensity to be measured in any direction in the whole spherical space around the luminaire; if there is an axis of symmetry the results can be averaged to yield the polar curve of light distribution but if there is no such symmetry the results can be plotted in the form of an iso-candela diagram which gives the whole information.

Directional measurements are often the starting point for a series of calculations for which tabulated figures are more useful than curves and diagrams. The calculations generally serve to reduce the directional data in such a way as to give a single figure, such as the total light flux in the lower hemisphere, or a series of figures such as the utilization factors for different room sizes. These results could generally be obtained by direct measurement but this would be less informative and sometimes less accurate.

We should perhaps mention a non-directional measure of illumination known as the scalar illuminance, the quotient of the amount of light in lumens incident on the surface of a (small) sphere by the area of the surface of the sphere in square metres. This is important as a measure of the lighting of irregular objects rather than of flat objects. It is the proper measure for the lighting of a place where people meet each other or for the lighting of a car park, even though neither of these involve very irregular objects, and it is easier to measure than to calculate. The measuring device is usually a modified table tennis ball or its equivalent, containing one or more photocells so that it is equally sensitive in all directions.

Flux and luminance

Luminous flux is the beginning of the process of seeing, as for example the flux output of a lamp, and luminance is one of the last stages in the process because it is actually the contrasts in luminance that stimulate the eye. Total flux output from a lamp is best measured by operating it in an internally whitened box which is large enough for the dimensions of the lamp to have no effect; the inter-reflected light which falls on the surface of the box is substantially uniform at all points and is directly proportional to the total flux released. A simple measurement by a photocell screened from the direct light from the lamp will indicate the total flux, subject to calibration by a standard lamp to assess the amplifying effects of the multiple inter-reflections. This is of course a laboratory technique but it is in regular routine use in lamp production. The same technique may be used to measure the light output from a luminaire relative to that from its lamp, the "light output ratio".

Luminance is less frequently measured, perhaps because nobody expects to pay for it. Luminance meters are liable to be complex, with telescopic lens systems or visual matching of brightness and the results are difficult to interpret because of the readiness with which our eyes adapt to the prevailing luminance level. Luminance design is however one of the objectives of our more advanced designers and some means of quick and reliable measurement will doubtless become as familiar as the light meter for measuring illuminance; we may expect to use an optical system so that we can see what object is being measured and to have a photocell to indicate the true luminance of the object.

Units, abbreviations and terms

The words which have been used in this chapter for photometric quantities are sometimes awkward and clumsy but they are precise

—very necessary in any engineering measurement. A list follows, with colloquial definitions:

Lumen (lm). Unit of *luminous flux*, the amount of luminous power or the total amount of light.

Lux (lx). Unit of *illuminance*, the amount of light (lumens) falling on unit area (square metre).

Candela (cd). Unit of *intensity* (candle power), the directional luminous power or the ability of a light source to produce illuminance at a distance.

Candela per square metre (cd/m^2 or nit). Unit of *luminance* (which is still sometimes called photometric brightness), the amount of directional light emitted or reflected from unit area of a surface.

Apostilb (asb). Another unit of *luminance*, broadly equivalent to the total amount of light reflected from unit area of a surface; strictly equal to the luminance of a perfectly diffusing (white) surface emitting one lumen per square metre. $1 \text{ cd}/m^2 = 3\cdot14$ asb.

Reflectance. Ratio of total reflected light flux to total incident light flux (previously known as reflection factor).

Proper definitions and quantities are given in British Standard 4727* for those who want the precise meanings. It is general practice in the lighting industry to use SI units and the habit of measuring distances in metres rather than feet or inches is spreading to all those concerned with the lighting of buildings as metrication develops.

Finally, a few notes on the "iso-candela diagram" which sometimes seems to immobilize the minds of otherwise perceptive people. It is quite simple, being only a contour map. The difficulty arises from the background grid; an ordinary contour map such as the one inch Ordnance Survey has a grid of north-south and east-west lines which represents real positions on the surface of the earth, but the grid of an iso-candela contour map is a series

---

*British Standard 4727, Part 4, Groups 01, 02 and 03 (1971-72)

of lines which represents directions in space, the directions of rays of light from the source. Fortunately there is a convenient comparison with the terrestrial globe with which we are all familiar. The grid of the usual iso-candela diagram is the same as the lines of latitude and longitude on a sphere surrounding the light source; the direction of any ray from the central light source is identified by the latitude and longitude of the point where it strikes the sphere. If we imagine the height of the land above sea level on the sphere to be proportional to the intensity of the ray of light, so that we have mountain ranges and plains corresponding to beams of high intensity and areas of uniform intensity, and if we draw contours of height, we have the iso-candela diagram of the light source on the surface of the sphere.

Traditionally, we project this spherical surface on to a plane in the form of an "onion diagram" whose proper name is a sinusoidal or Sanson–Flamsteed projection. This is shown in Fig. 3.1. The

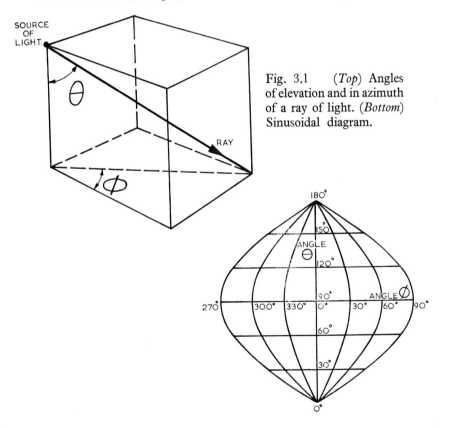

Fig. 3.1 (*Top*) Angles of elevation and in azimuth of a ray of light. (*Bottom*) Sinusoidal diagram.

top sketch shows the angle of elevation θ and the angle in azimuth ϕ of a ray of light according to the usual convention. These angles are sometimes designated γ and C respectively. The lower sketch shows a sinusoidal projection of a hemisphere with scales of θ and ϕ on the lines of latitude and longitude; this is an "equal area" projection in that it has the property that any square millimetre of the diagram represents the same amount of solid angle wherever it may be. This means that the product of area by intensity (which may be read from the contours) gives the luminous flux, just as the product of area and height on an ordinary map gives the volume of a mountain. No iso-candela contours are marked on this diagram because they would complicate the explanation!

There is a new tradition of circular iso-candela diagrams which are more convenient although less impressive than the onion shape. The diameter of the diagram is 1·414 times the diameter of the sphere which it represents, so that its total area is equal to that of a hemisphere ($2\pi r^2$). It can be projected in either of two ways as shown in Figs. 3.2 and 3.3. Imagine the sphere resting on a plane, with the south pole touching the plane (Fig. 3.2) or the equator touching the plane (Fig. 3.3). The latitude and longitude circles on the spheres are projected down on to the plane to form the diagrams shown in the lower sketches, at angles which have to be properly calculated to provide the equal area property. These are the grids for the circular iso-candela diagrams, with just the same lines of latitude and longitude and the same equal area property as the onion shape but with two other convenient features: the same beam projects into contours of the same size and shape on either diagram and the effect of tilting the light source, so as to swing the beam in the plane of the paper, can be represented by turning either diagram about its own centre. This means that a floodlight may be measured on a goniometer with one set of co-ordinates and the curves plotted very easily on a different set, or the calculations made and recorded on a different set.

For those who may wish to be precise, the proper names of these

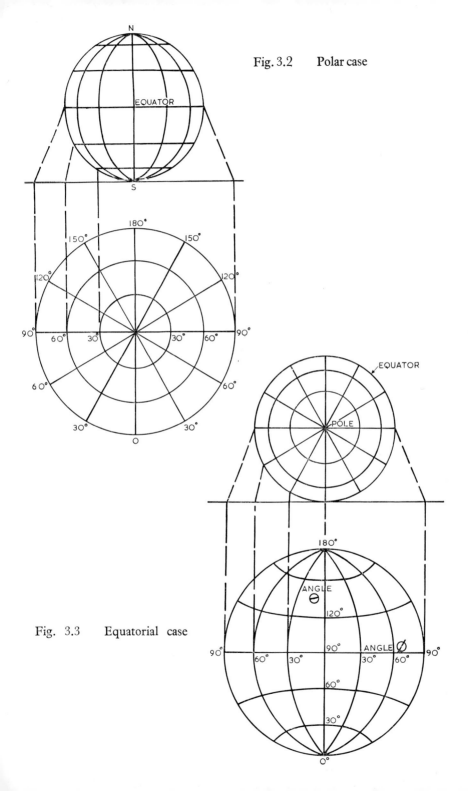

Fig. 3.2 Polar case

Fig. 3.3 Equatorial case

circular iso-candela grids are zenithal equal area projections, polar case and equatorial case respectively as shown. The numbering of the angular scales differs from the usual angles of latitude and longitude but the lines are the same.

There can of course be other forms of iso-candela diagram, such as a rectangular grid for a segment of a sphere or an oblique projection or a specially convenient one which looks like a mixture of the two lower circular sketches and which has grid lines corresponding to the kerb lines and transverse lines in street lighting. All these have the same basic principle; contours of constant luminous intensity are drawn on a grid representing angular directions in the space around the light source. The great advantage of an iso-candela diagram is that it provides a precise pictorial record, on one piece of paper, of the whole information about the distribution of intensity from a light source or projector.

To complete this rather complex portrayal of light distributions, there may be iso-illuminance curves, usually known as isolux or isophot diagrams, or iso-luminance curves. These are usually drawn on grids representing actual positions on the surface being illuminated rather than on angular scales; they are just like ordinary contour maps of illuminance or luminance.

4 Dirt and Maintenance

Electric light, which costs real money, is destroyed by dirt before ever it can be used. That one brief sentence is the condemnation of a dirty lighting installation. Who ever saw a dirty lens in a lighthouse?

Dust and dirt are anathema to the lighting engineer because his instinct and training have given him a reverence for light. To the accountant, dirt and maintenance are a nicely balanced exercise in cost effectiveness, offsetting the real costs of cleaning against what he calls the notional value of the light that is lost. To the rest of us, dirt is a double nuisance, making things look grimy and causing the bother of cleaning or redecoration.

Where does the dirt come from? It is nearly all airborne, by ventilation from the outside or from processes causing the degradation of materials inside; fumes, oil droplets and greasy vapours make it much worse. Relatively little dirt enters on shoes or clothing if ordinary standards are observed, but the wearing-out of carpets or cloth or concrete can be a continuing source of dust.

Why does a luminaire get dirty, when it seems to be suspended out of harm's way? Partly because it is warm and partly because the air around it is moving slowly; the laws of physics, electrostatic attraction, the van der Waals adhesion forces, apply to all surfaces whether plastics, metal, glass or paint. Warm air will deposit dirt on cooler surfaces; hygroscopic surfaces (such as glass) collect dirt in a humid atmosphere; non-hygroscopic surfaces (such as plastics) attract dust in a dry atmosphere; coated surfaces (such as anodized aluminium) can develop strong charges; the chemical action of corrosion (on steel or aluminium) roughens the surface. There seem to be no physical laws operating in room interiors which would loosen the adhesion of dirt, particularly of greasy dirt. A rapid flow of cool air may dislodge dust or at least will not deposit the particles carried in suspension; this scouring effect is the justification for slotted reflectors and is one of the merits of air handling luminaires.

Cleaning is both a chemical and a physical process; chemical in softening the agents which provide adhesion and physical in carrying the dirt away in water, air or fabric. Some surfaces such as glass or stainless steel can be cleaned more easily and more vigorously than others, such as paint or aluminium. Generous washing in a weak, warm detergent solution ought to be enough—any luminaires which require steel wool or brick dust or solvent cleaners ought to have been cleaned before—and even materials with non-wettable surfaces such as some plastics can usually be cleaned with a somewhat stronger detergent which will wet the dirt particles and wash them away.

The effect of dirt

Dirty luminaires or dirty room surfaces eliminate some of the sparkle and make a place look dingy. In photometric terms, dirt reduces the spread of light and therefore increases the diversity of illuminance; it reduces inter-reflection and therefore makes the lighting more direct. There is often a greater loss of light than would be guessed—if an installation is just beginning to look dirty, at least 20 per cent increase can be expected after thorough cleaning; if the whole room also looks dirty, the effect of cleaning and redecoration will be nearer to 100 per cent increase in luminances. All the advances in technology of lamps and of light control, all the care in design and manufacture and all the detailed specifications of the installation may be nullified by dirt.

The design of the luminaire is an important factor. The simplest reflector lamp is probably least affected by dirt because it has no external light-controlling surfaces, but it is not generally good lighting and moreover any grime which does get burnt on to the bulb is hard to remove. Open reflectors with ventilation slots and dust-tight enclosures resist the deposition of dirt longer than either louvred luminaires or unsealed enclosures; an enclosing dish or diffuser below the lamp may seem to be a protection against dirt but it is more likely to act as a dust trap, carrying a layer of unseen dirt on its upper surface. All enclosed luminaires breathe in when switched off, taking in dust which is not breathed out when the lamp is switched on again; gasket materials should therefore be chosen to act as a dust filter when the cover is closed. Prismatic diffusers should have their prisms on the lower surface. Air handling luminaires are usually designed to avoid serious deposition of dirt inside them but if warm air flows past a cool surface it generally leaves a dirty trail.

Indirect lighting is particularly sensitive to depreciation, partly because the upward-facing surfaces of the luminaires collect dust which cannot be seen and partly because the ceiling surface suffers the deposition of dirt from the streams of warm air rising from the lamps at the very spots where the ceiling is most brightly lit.

Dirt interferes with daylight as well as with the view through a window. Daylight is not free because the windows involve capital costs, heating costs and maintenance, so dirty windows are a waste of real money. Apart from this, daylight is precious in many interiors though they may be lit by electricity.

Dirty windows, dirty luminaires and dirty lamps seem to be more objectionable than dirty ceilings, walls or furniture. They imply neglect and failure to appreciate the importance of seeing and of enjoying what we see. We have a natural instinct that light and dirt do not mix.

The cost of dirt

Dirt absorbs light which has cost real money and which was generated in lamps and equipment which have cost real money. This matter is discussed in detail in IES Technical Report No 9.

The "maintenance factor" of an installation is the ratio of the amount of useful light available after a specified period of use, to the amount from a new, clean installation. Sometimes the "average maintenance factor" may be stated for the averaged amount of useful light available during a period of use.

It is interesting that the average maintenance factor also indicates the proportion of the total cost of the lighting installation that is lost by dirt and by physical depreciation.

The maintenance factor to be used in a design must be chosen carefully to suit the operating conditions. It ought always to be stated, no less than the objective illuminance, to emphasize that the designed value will only be achieved in service if the maintenance arrangements are in line with the factor chosen. In particular, the design maintenance factor should be declared in tender documents.

The cost of lamp replacement

The whole cost of a new lamp comprises the price from the supplier, the administrative costs of buying it and paying for it, the cost of storing it and the cost of taking out the old lamp and inserting the new one. Whether lamp replacements should be on an individual fail–renew basis or on a bulk replacement basis depends largely on the size of the installation and on the organization of the labour force used for maintenance. The optimum economic choice depends on too many factors to be examined in this short chapter but it may be said that, except in an organization where the electrician is standing by and doing nothing much else, the cheaper and more satisfactory arrangement for an installation of more than a score of lamps is bulk replacement.

Tungsten filament lamps generally diminish about 10 per cent in light output and then go out; there is never much doubt about the time for renewing them. Discharge lamps, including the fluorescent bulb types, will continue to emit some light long after their economic life is over; the light output may drop to 80 per cent after say 5000 hours, but the lamps may go on and on, giving only half their initial output after say 15 000 hours but still consuming the same electrical power. The same is true of tubular fluorescent lamps.

Lamp replacement is therefore a matter for discipline and firm resolution. It is not easy to destroy a lamp while it still appears to work, even though its light output is reduced and may cost half as much again as that from a new lamp. When the cost of the light which will *not* be generated during the remaining burning hours of a lamp because of its depreciation exceeds the cost of a new lamp, it pays to change the lamp irrespective of the illuminance. Bulk replacement involves this decision every time but some relief may be gained by a simple record system which can enable occasional failures to be replaced by part-used rather than by new lamps.

Maintenance

There are two kinds of maintenance—essential and optional. For the first, the renewal of lamps, ballasts or fuses and the repair of switches or of any electrical or mechanical devices in a dangerous condition are necessary if the installation is to continue as a source of light, whatever its quality; the figures for this essential maintenance should be taken into the costing of the operation of the installation as an on-going charge. Emergency lighting comes into this category. For the second, the cleaning of luminaires, ceilings or walls and the replacement or re-painting of uncleanable surfaces are optional in that the installation is not inhibited from producing light of a kind if this work is not done; the figures for optional maintenance are also part of the on-going costs but they depend on the economics of the light destroyed and of the amenity which is reduced by the dirty conditions.

Essential maintenance requires regular inspection and, as indicated above, a decision on either fail–renew or bulk replacement of lamps. Optional maintenance requires selection of the right techniques of cleaning to suit the circumstances of the installation; these are reviewed in IES Technical Report No 9.

The occasion of lamp changing is commonly chosen as a time to clean the luminaire; if this is done on a fail–renew basis, some of the luminaires will be much brighter than others, and vice versa. The cleaning period is, or should be, shorter than the lamp life so there will usually be several cleanings between the lamp renewals.

The cost of maintenance

A lighting installation is perhaps the only situation in which the cost of dirt can be assessed accurately and related to the cost of cleaning. The operating costs are the same, whether it is clean or dirty, but more light can be obtained just for the cost of cleaning. The frequency of cleaning has to be balanced with the increased light output and the whole complicated story is worked out in IES Monograph No 9 and IES Technical Report No 9.

If economics are more important than amenity the only factors involved are the total annual operating cost of the installation, the rate of depreciation and the cost of cleaning. The cheapest arrangement is to clean each luminaire at the time when the value of the light lost is equal to the cost of cleaning it; if it were left longer, the value of the wasted light would exceed the cost of cleaning, which would be to nobody's benefit.

We can explain it best with some simple mathematics. D is the annual rate of depreciation (the relative loss of useful light per year) and C is the ratio of the cost of one cleaning of the installation to the total annual cost of the installation. Then the "economic cleaning interval" is given by:

$$T \text{ (in years)} = \left(\frac{2C}{D}\right)^{\frac{1}{2}} - C$$

This is over-simplified, but Technical Report No 9 is not easy to understand and some simplification is justifiable.

The values of the depreciation rate D for a good quality enclosed luminaire may be taken from the Technical Report as in Table 4.1.

Table 4.1

Environment	Annual rate of depreciation D
Air conditioned office	0·05
Country office	0·15
Suburban office	0·25
Factory office	0·4
Machine shop	0·6
Foundry	0·8

The formula is awkward to calculate so the figures in Table 4.2 are given for average and bad rates of depreciation, namely 30 per cent and 70 per cent.

At the bottom of Table 4.2, for high relative cost of cleaning, the light output just before cleaning would be about half the design value or less, which implies a great sacrifice of amenity to economics. As in many things, it is cheaper to leave things dirty than to clean them but nobody would regard this as desirable.

Table 4.2

Cost of one cleaning ÷ total annual running cost	Economic cleaning interval $D = 0.3$	$D = 0.7$
0·01	3 months	2 months
0·02	4	$2\tfrac{1}{2}$
0·05	6	4
0·1	8	5
0·15	10	6
0·2	11	$6\tfrac{1}{2}$
0·3	13	$7\tfrac{1}{2}$
0·4	15	8
0·5	16	8

Does it pay to over-light and then to clean less frequently? This is sometimes done when cleaning is difficult and would interfere with production processes; in extreme cases, such as an airway obstruction light on a high chimney needing a steeple jack for lamp renewal, it is obviously good economic sense. In ordinary lighting situations it is a matter for debate, taking account of the great change which occurs when cleaning is done. For example, an installation which requires annual lamp renewal and has rapid depreciation (say 60 per cent per annum) might be designed with its wattage increased to 50 per cent above its normal design value and then it could be cleaned and re-lamped once a year; the working illuminance below the installation would be suddenly doubled or more once a year, which might be bad for morale. Fortunately, this technique of over-lighting is rare.

This discussion of dust and dirt and the expense of removing them is a rather negative and depressing exercise. To take a more positive view, we may remember that light is necessary to our enjoyment of our visual environment and that the word "maintenance" (from the French *main, tenir* and the Latin *manu, tenere*) means "to hold in the hand", to conserve in the same state, to preserve the amenity. Because something is worth having, it is worth looking after, it is worth some effort and expense to maintain it against depreciation. A detailed maintenance schedule for a lighting installation is more than a cost effective elimination of wastage or the avoidance of complaint; it is a policy of preserving the quality of the lighting as an essential amenity in our surroundings. If the user is to have the light he wants, the standard of maintenance is no less important than the standard of lighting.

5 Colour

Colour is a sensation, perceived through the light-sensitive receptors of the human eye. There must therefore be light and an observer before colour in this sense can exist. The light may come direct from a source itself or it may come by reflection or transmission from some object which modifies it (and we may then say that the object is "coloured"). The eye receives this light and transforms it into impulses in the retina and in the optic nerve; the brain has to translate these impulses into sensations and, by reference to stored memory, to decide what colour is being "seen" before we can say that the colour is "perceived". We all of us "see" an extremely complicated picture all our waking hours but we only "perceive" when the brain takes notice and brings its analysis of the impulses into our conscious minds. We thus have a train of agents:

1. Light.
2. An object which may be the light source or something which modifies the light.
3. A seeing eye.
4. A brain to interpret.

Light is not only the essential initiator of this series but it contributes vitally to exactly what we see. Our interpretation depends on our phenomenal ability to remember colour as a distinct sensation, independently of the light or the object which appears to be coloured.

Coloured light

Our natural illuminant is daylight from the sun. This has a continuous spectrum, i.e. it contains light of every wavelength within the visible range from 380 to 760 nm, and these all combine to form what we call white light. The fact that this is really a mixture is shown by splitting it into its component parts (spectrum colours) with a prism; this experiment demonstrates that these component wavelengths when separated give a sensation of colour which is quite different from the sensation of whiteness. There is a remarkable consistency amongst people from all parts of the world in the perception of colours, so much so that there is no noticeable difference amongst 95 per cent of the population and it is possible to calculate or measure colours to a real accuracy of one part in a thousand.

The simplest explanation of colour vision (although this is constantly being assailed) is that different mechanisms in the eye and its associated nerve centres react to different wavelengths, rather in the way that radio receivers can be tuned to stations of different wavelengths but play the programmes through one loudspeaker. A satisfactory theory of colour perception can be built on three types of receptor which are chiefly sensitive to the long wavelengths (the red end of the spectrum), the short wavelengths (blue and violet light) and the middle part (chiefly green light). By mixing colours of this description, every colour sensation can be produced. All three in equal balance produce a sensation which we call white, even though a continuous spectrum is not present; reduction in any one of these components gives a sensation of colour and complete absence of light gives us black. Stage lighting is often designed on this principle.

A light source may therefore be seen as coloured if it has a discontinuous spectrum with serious gaps, as in neon, sodium or mercury lamps. Tungsten lamps give a continuous spectrum which, although emphasizing the red and yellow part of the spectrum, is acceptable as a warm, white light. Tubular fluorescent lamps give a spectrum in which the gaps are filled by the light from the fluorescent powder, emphasizing the red, green or blue depending on the constitution of the powder, and are also accepted as white lights even though they are not all the same.

The colours of incandescent lamps can be specified by stating their "colour temperature" which is the temperature of a non-selectively radiating light source of the same colour. The ideal non-selective radiator is a black body and the standard is a carbon tube externally heated to incandescence; the colour and the spectral quality of the light from the inside surface are fully described by its temperature on the Kelvin scale. The convenience of this method of description of the appearance of a lamp has led to the term "correlated colour temperature" for the colour appearance of other light sources having a near-continuous spectrum; this may be found by mathematical computations of a complex type but it is no more than an approximation. The colour appearance of discontinuous spectrum sources, and all non-emitting objects such as surface colours or translucent materials, cannot be specified in this way.

Coloured things

We are more interested in the colours of objects around us than in the colours of light sources; we associate the colour with the object because it modifies the light by absorbing some part of its spectrum. If daylight is reflected from a "red" object, the colour sensation is caused by the object having the power to absorb the green light falling on it and to reflect the rest of the light. If it absorbed only the green, it would reflect red and blue and we should see a crimson colour; if it absorbed green and blue we should see a pure red. This power of selective reflection is such a prominent and usually permanent property of an object that we say it has a crimson or red colour, whatever the properties of the light by which we see the object, and we sometimes grumble if the "colour of the object" seems to change in different lights. In fact, a coloured object owes its colour to its ability to absorb some of the light falling on it. It is puzzling to artists to tell them that they are not adding colour when they are painting a picture; they are adding pigment, which absorbs some of the incident light; what we see is the light which is reflected and not absorbed by the pigment.

An object cannot reflect light of wavelengths which are not present in the incident light, so it will "change colour" if viewed in a light deficient in that part of the spectrum which it does not absorb. Thus if meat displayed on a butcher's counter is seen under fluorescent lamps deficient in the red part of the spectrum, it will look an unappetizing brownish grey because there is insufficient red for it to reflect. Similarly, flesh colours may suffer but salads would look fresher because the green components in the light would be a high proportion of the total reflected light.

The eye–brain combination is very adaptable and we soon get used to our surroundings. If there is some surface in a room which is known to be white—i.e. non-absorbing, and we call it "white"—then after a short period of adaptation we accept all the colours in the room as being normal unless the illumination is very far off white or we can see through a door or window into another area which has white lighting. Incidentally, if there is no white or true grey surface to be seen in our surroundings, our ideas as to the true colours may become very confused.

Metamerism is a troublesome phenomenon of colour perception which may be difficult to understand. Two coloured materials which have different reflectance characteristics throughout the spectrum may appear to be of the same colour when seen under some particular illuminant because the parts of the incident light which they reflect produce the same sensation; under another illuminant with a different energy distribution however, they will no longer appear to match. This effect does not occur when the materials are prepared from identical pigments or dyes but the trouble may be unavoidable if there is a variety of colourants, such as in a bathroom where there may be porcelain, enamelled iron, plastics, cottons and wool which may all match in daylight but look entirely different by electric light. The natural adaptation of the eyes reduces this effect with continuous spectrum sources but may not operate with partially discontinuous spectrum sources such as some high efficacy lamps. Before passing a colour ensemble in any room, it is desirable to view it under all of the possible illuminations that may be employed, unless of course one is looking for a bizarre effect.

Simultaneous colour contrast can also modify the sensation of colour. If there are two or more colours in an area directly under observation, the way in which the brain interprets the colours is modified, sometimes diminishing the differences by a spreading or diffusing effect and sometimes enhancing them. A bright colour seen against a grey background will appear brighter and the grey will tend to take on the complementary hue, probably owing to a short-term fatigue effect. Two bright colours of different hue in close juxtaposition will have the effect of dulling each other or of making the hues change. Thus in a highly coloured design in fabric or in printing, the individual colours may change considerably in appearance and it is possible, by making changes in the design or in the background, to obtain the effect of using many colours from only a few carefully selected basic colours.

A different phenomenon arises from the unavoidable and entirely natural aberrations in our eyes by which we focus differently with lights of different wavelength. A bright red patch on a grey background may appear to be nearer to us than its background, or bright blue-green lettering on a red background may be almost impossible to read.

Colour appearance and colour rendering

Colour appearance is what a lamp looks like; colour rendering is what it makes other things look like. Tubular fluorescent lamps are available in many different "white" colours and we may sometimes blame careless maintenance men for getting them mixed, but they are also available with different colour rendering properties which are less obvious to the unskilled eye but no less disastrous if they are mixed.

Poor colour rendering means that the colours of familiar objects appear to be distorted, even after a period of adaptation, and the degree of distortion cannot be properly expressed by a single figure. The distortion may be due to the lighting being "warm" or "cool" ("Colours seen by candle light do not look the same by day.") or, more troublesome, to the spectral composition of the light being irregular. It is a complex problem because there are adaptation and familiarity effects which are almost impossible to assess and because even similar looking colours may be distorted differently. A single figure of merit for a lamp may be obtained from Dr Crawford's 6-band method and a colour rendering index is given by the CIE colour shift method, but both are averages and neither gives any precise information about particular situations. Colour rendering of lamps certainly varies from bad, through poor and medium to good but the verbal description of a lamp depends on what it is used for; we may say that one lamp is better than another without being sure that either is good. Informed experience seems to be a better guide than calculation or specification in our present state of knowledge, but our colorimetrists are working on it.

Defective colour vision

Colour blindness is very rare indeed. Less than 5 per cent of the population differs to any noticeable extent from the average in their ability to distinguish different colours and most of these manage without any trouble at all—there is no stigma in defective colour vision and very little disadvantage unless you want to be a pilot or a dyer, apart from the loss of one of the pleasures of life.

The usual trouble is confusion between pale colours of the orange, lemon or yellow-green hues or between brown, khaki and olive; there may be decreased sensitivity to the long wavelength (deep red) light in one or two per cent of males. The total incidence of these defects is about 8 per cent of males and $0 \cdot 4$ per cent of females. A rare defect is confusion between blue and green or between white and yellow when seen as large coloured areas; we all have this defect to a slight extent in the very centre of our vision (known to the experts as foveal tritanopia) when we look directly at a small coloured spot but we are so used to it that it does not affect us seriously.

There is no adequate explanation of defective colour vision and no cure nor even any remedy such as special training or tinted glasses; the sensation of colour seems to be too deeply rooted in memory and instinct for any innate confusion to be resolved. Defective colour vision is congenital, largely hereditary, and is not acquired or changed during life except rarely by accidental damage to the eye.

Description and specification

The human eye can distinguish several million different colours. Precise description is essential in manufacture and commerce; verbal descriptions are wholly inadequate. On the other hand, our impressions of colour can only be communicated in words, so the precise system must be understandable in verbal thinking if it is to be readily used. Colorimetrists have evolved an international numerical system which works extraordinarily well, defining mixtures of precisely known coloured lights which combine to match the colour sensation of the sample, subject of course to prescribed conditions of illumination and measurement.

The body which sponsored this basic system of defining the appearance of a colour is the Commission Internationale de l'Eclairage (CIE) through its Colorimetry Committee, on which this country has always been strongly represented. The original agreement was reached in 1931 since when additions have been made without altering the original concept but now occupying 124 pages in CIE Report 15.* The CIE system expresses its result for a coloured article in terms of x and y (proportions of idealized red and green) and of Y (total light or reflectance) for a stated type of illumination; the x and y co-ordinates may be plotted on a CIE chromaticity chart to form a diagrammatic representation of the actual colour.

The Munsell system for coloured surfaces is aligned with the CIE system but includes the provision of standardized colour chips which are calibrated in three scales:

Hue The kind of colour, red, yellow, etc.

Chroma The purity or saturation of the hue, analogous to the dyer's "strength". A difficult concept but fortunately the chroma scale is precise.

Value The lightness of the colour, linked with reflectance.

In the recent BS document DD 17, "greyness" is used as the inverse of chroma and "weight" is approximately correlated with value. Any coloured articles having the same three parameters in the Munsell system will look the same colour whatever the lighting conditions, subject to any metameric difficulties.

The scale figures in these two systems, for light reflected from

---

*CIE publications are supplied through the National Illumination Committee of Great Britain, Building Research Station, Watford.

an article in the CIE system or for the colour properties of an article in the Munsell system, require accurate measurement or complex calculations using the defined colour response of a "standard observer", the spectral reflecting or transmitting properties of the article and the energy distribution of the illuminant. There are four standard illuminants for scientific work:

Illuminant A   tungsten lamp (colour temperature 2856 K).
Illuminant B   represents sunlight (4900 K approx.).
Illuminant C   represents north sky (6900 K approx.).
Illuminant D_{65} average daylight with some ultraviolet.

In any system, a coloured article may be specified by the numerical description, with appropriate tolerances.

Measurement of colour

Increasing precision has involved increasing sophistication, such that most colour measurements nowadays are made in grey boxes containing automatic photoelectric spectrophotometers with computer linked printout. For simpler minds, there are two photoelectric methods and two visual methods, apart from looking through a colour card or an atlas of Munsell chips and trying to find the nearest match. Photoelectrically, one may measure the energy distribution right through the spectrum and calculate the answer or, less accurately, measure the energy in three portions of the spectrum chosen to yield the x, y and Y values in the CIE system almost directly. Visually, one may use an additive (Hilger–Guild or Donaldson type) colorimeter to match the test colour by physical addition of red, green and blue lights, which requires a well-trained observer to ensure accuracy, or a subtractive (Tintometer) colorimeter in which actual primary filters are used to build up the same colour as the test colour, requiring an observer of normal colour vision.

Whatever instrument is used, one has to decide whether to measure an absolute value or a difference from some agreed standard; the former is difficult and expensive whilst in many cases the comparative method is more accurate, easier and cheaper. Before embarking on any form of colorimetry, it is wise to consult an expert in one's own particular field and to decide which approach is most appropriate.

There is no doubt that some understanding of colour theory is essential before designing an interior lighting decor which will strike the right note under all conditions. This is exemplified by the high proportion of lighting engineers in the membership of the Colour Group of Great Britain.

6 Light Control

The International Lighting Vocabulary (1970) gives the definition:

Lighting fitting (UK); *luminaire* (USA): Apparatus which distributes, filters or transforms the light given by a lamp or lamps and which includes all the items necessary for fixing and protecting these lamps and for connecting them to the supply circuit.

This definition is published by the Commission Internationale de l'Eclairage (CIE) and by the International Electrotechnical Commission (IEC). The term "luminaire" is increasingly used in this country as well as in the USA and elsewhere.

Considering that the essence of a lighting fitting is usually the optical system, and that the basic means for distributing the light are so few and so simple, it is perhaps surprising that there are so many different fittings on sale and that so many new ones keep making an appearance. The emphasis on glare control and on efficient distribution of the light from lamps within the BZ system has made careful optical design important for all except purely decorative luminaires.

Control of light by reflectors

There are three commonly used methods of light control, namely reflection, refraction and absorption; we learned the fundamentals of the first at school. The most interesting and powerful form is the mirror type (specular) which obeys two laws: (1) the angle of reflection equals the angle of incidence and (2) the incident ray, the reflected ray and the normal to the mirror lie in the same plane. Angles of incidence and reflection are almost invariably measured with respect to the normal (i.e. the line at right angles) to the reflecting surface.

As we all know, mirrors can be shaped so that they form an image of an object and by applying the laws which control the reflections we can calculate exactly where the image is, how large it is and so on. For example, in the simple case of a flat looking glass we see an image of ourselves, the same size as ourself, as far behind the mirror as we are in front. We can make a curved mirror which will produce a larger or smaller image, in front or behind; in amusement arcades the curved mirrors distort the curves of the viewers in ways that may be amusing or pathetic or flattering or unkind, depending on whose image we look at. If we put a light source in front of a curved mirror, we can calculate the distribution of light which will be reflected from the mirror. A simple and well-known example is the searchlight which concentrates into a very narrow angle beam the light which emanates from its focal point. The contour of the searchlight mirror is a parabola, or rather a paraboloid of revolution based on that focal point.

Other shaped mirrors may spread light rather than concentrate it—although concentration in some direction or another is the most usual requirement—and we can calculate the light distribution. Working in reverse, we can determine the shape of mirror which will produce a given desired distribution. The distributions so obtained may be widely different. As examples, we may produce the narrow pencil beam of the spotlight or the complex double peak of the street lighting lantern; we may finish with a dish shaped reflector, symmetrical about an axis of revolution or with one having symmetry about a plane; we may use trough-shaped reflectors; we may have perfectly smooth surfaces or we may impose flutes to modify the distribution of the main reflecting contour. All these very different shapes derive from the two basic laws of specular (or regular) reflection.

Another commonly-used reflecting surface is the preferential reflector which has slightly diffusing properties. Aluminium, for example, may be slightly de-polished so that it reflects a bundle of rays grouped around a given direction but without forming a clear image of the object. Such a surface is very useful. Sometimes the diffusion is used to smooth out small irregularities and inaccuracies of manufacture; sometimes, for example where different parts of an optical system have to work together, diffusion can help run the one distribution into the other. The amount of diffusion employed may be critical and a proper understanding of preferential reflection is an important part of the design process.

The most widely-used reflector is the complete diffuser such as white stove enamel and vitreous enamel. The theoretical concept is a surface which looks equally bright from any angle of view, and whatever the angle of the incident light; good approximations are white blotting paper, snow or matt white paint. Light reflected from such a surface is distributed uniformly in all directions, so that the luminous intensity emitted in a given direction is proportional to the area facing that direction. If the surface is a flat plate, the maximum intensity is at right angles to the plate and the intensity at any other angle of view is proportional to the trigonometric cosine of that angle (because the projected area is proportional to the cosine of the angle). When this is worked out, we find that the polar curve of light distribution is circular with a maximum intensity at 0 degrees and zero at 90 degrees (BZ Classification 5).

If a diffusing reflector of a lighting fitting has a flat mouth, this is optically equivalent to a flat disc filling that mouth, from which it follows that most diffusing fittings give a symmetrical light distribution from the reflector, with the axis of symmetry normal to the mouth aperture and with a circular polar curve. To this must be added the distribution of the bare lamp to get the overall light distribution. The diffusing reflector is thus very limited in the light control that it can effect but fortunately it so happens that this distribution is of great practical value and is therefore widely used. The old conical opal glass shade, the more sophisticated dispersive reflectors for incandescent lamps and the open trough reflectors for fluorescent lamps are important examples.

In fact, however, none of these uses a genuinely uniform diffuser because the practical surface has an overlaying glaze which gives a smooth, easily cleaned finish. There is therefore a fourth class of reflector which is a diffuser with a specular surface imposed on it; the specular reflectance is generally about 5 per cent and does not greatly alter the light distribution produced by the diffusing element, which may have a reflectance greater than 75 per cent, but it can have benefits in design and should not be ignored. This mixed type of reflection is very common and nearly all surfaces, however matt they may appear at normal angles of view, will show specular reflection at glancing angles.

Control of light by refractors

Refractors are hardly less important than reflectors and are in many ways more intriguing; the process of refraction occurs in prismatic panels or lenses or preferential diffusers. Again, two simple rules control the bending of the rays passing from one transparent medium to another: (1) the sine of the angle of incidence equals the sine of the angle of refraction multiplied by a constant and (2) the incident ray, the refracted ray and the normal to the refracting boundary lie in the same plane. The constant is called the refractive index: for glass and for the transparent plastics materials in general use it is about 1·5.

The most familiar everyday use of the phenomenon of refraction is the lens. Modern lens design is highly developed and the most beautiful and precise images can be formed. The microscope makes images of objects almost touching the lens; the telescope stretches light years into space.

For the fittings designer, the prism is more useful than the lens. Prismatic formations can be made to cover large areas and therefore to pick up a large fraction of the light emitted from the source more efficiently than lenses. Prismatic plates are also economical in use of material because there is no thick central section, as is inevitable with a lens. A prism can be designed to bend light through a small or large angle. It can even act as a reflector, turning the light back upon itself like a mirror. By putting curved faces on the prisms a controlled spread can be impressed on the light passing through, giving another freedom of design. Sometimes, but less often, a diffusing layer or pattern is also applied to the surface.

One great advantage of prismatic control is that the overall contour is much less bound by the required light distribution than is the case for specular reflectors. For example, to achieve a narrow beam with a mirror, the contour must be parabolic—this is inevitable. But with prismatic control, the general shape of the refracting component may be designed first to suit the mechanical or aesthetic requirements and prisms of the necessary angles for light control can then be formed on that shape. Another real advantage is that the prismatic form is in some respects less sensitive to errors in manufacture or assembly; whilst an error of 1° in the inclination of a mirror will deviate the reflected ray by 2°, an error of 1° in the refracting angle of a prism will usually disturb the refracted ray by less than 0·5°.

The refractor was for many years the principal optical component used in road lighting lanterns in this country. The old high angle beams with their maximum intensity at about 80° from the downward vertical could be achieved most efficiently with refractors, of which there were elegant and complex designs using pressed heat resisting glass. The present day preference for semi-cut-off or cut-off lighting has changed the situation and requires a lower beam angle and a sharp run-back above the peak, which makes it more advantageous to use pressed metal specular reflectors.

For interior lighting, the large size of fluorescent tubes must be matched with large optical components. Fortunately, these lamps generate little heat in comparison with their size so that modern plastics are quite capable of withstanding the operating temperatures. Their capability of being accurately injection moulded into prismatic forms has led to a wide variety of shapes and performances, for most of which the design aim is to increase the downward concentration of light and to reduce the sideways intensities so as to avoid discomfort glare. If prisms of the correct angles and shape are formed on the surface of the refractor panel, they perform two functions: to bend the light nearer to the downward vertical, and to act as reflecting elements for light which would otherwise emerge near the horizontal. For this high angle light, which reaches the prism from the more remote parts of the lamps, the prism acts by total internal reflection without any absorption; it returns this unwanted light into the fitting where it is reflected from the interior surfaces and has a subsequent chance of emerging in a useful direction. This optical "louvre" is an interesting example of the ingenuity which designers bring to their craft.

White or colourless refractors and diffusers, unlike reflectors, do not absorb light to any significant extent, usually less than 2 or 3 per cent. They do, however, reflect part of the light which falls on them and this reflected light may suffer partial absorption in the fitting. Again there is scope for ingenuity in reducing these losses and achieving a high light output ratio.

Control of light by diffusers

Another important example of the control of light by transmission is the translucent opal diffuser. In opal glass or plastics materials, minute particles of a transparent substance are embedded in a matrix of another transparent substance of different refractive index; the scattering of light passing through can give excellent diffusion within the body of the material, with very low absorption of the light. The various opals, together with a wide range of patterned finishes both in glass and in plastics, provide a considerable variety of diffusing transmitting materials.

Control of light by absorption

The control of unwanted light by absorption, particularly in combination with reflection and refraction, calls for great skill by the designer. In all luminaires there are boundaries where the light is cut off by opaque areas, as in the simplest form of "shade" which was intended early in the history of design to reduce glare and to be combined with a reflector conserving the light.

We are now familiar with obfuscation in more elaborate forms, particularly as louvres. We find concentric louvres on projectors, "egg crate" and other opaque screening devices on fluorescent tube luminaires and so on, each designed to prevent light from going where it is not wanted. In the extreme case we sacrifice the light if we must; preferably we redirect it elsewhere. Thus louvres may be painted white; being opaque, they cut off the view of a bare lamp whose luminance would be unacceptably high and, being white reflectors, they redirect the light in other downward directions at a tolerable luminance. Similarly, louvres may be of an opal translucent material providing an effective cut-off without fully obstructing the light.

There are also those most interesting louvres which act as specular or preferential reflectors. In one form, horizontal reflecting "prisms" are formed on vertical aluminium louvres with the prism angles so chosen that all the light falling on them is reflected downwards; in another form, aluminized wedge-shaped plastics louvres have a parabolic section such that no light can be reflected above 45°. These louvre panels necessarily absorb some light but all the light they transmit is useful and they give a dramatically dark ceiling even in high illuminance installations.

Optical materials

The introduction in post-war years of plastics and of anodized aluminium to lighting fittings has greatly influenced designs of optical systems. Refractor plates of glass large enough for 5 ft fluorescent tubes have been made in smaller pieces which could be fixed together, but glass mouldings of say 150 cm by 30 cm are simply not practicable. In this sense, therefore, acrylic and polystyrene transparent plastics may claim to have made refractor optics possible for fluorescent lighting. Very sophisticated methods of manufacture are now available; injection moulding gives precise prismatic form in panels and troughs up to 2 m long; extrusion permits the combination of opal and clear prismatic plastics and in greater lengths; pressure and suction bending can form sharp edges on the shallow box-like enclosures which are now commonly used with fluorescent tubes. The products are surprisingly cheap but only when considerable capital outlay has provided the tools. Mass production of large numbers of a given design is generally essential to achieve reasonable prices.

Glass is still the medium of choice for use with the physically small, high wattage lamps such as the high pressure discharge lamps, since no transparent plastics competes in temperature resistance. For specular and preferential reflection, however, aluminium is now generally preferred to glass. It has a high reflectivity, is durable, will withstand high temperatures and can be formed into complicated shapes. As with plastics, the tool charges may be high but for long run production the overall cost per item is acceptable. For large specular reflectors, such as troughs for fluorescent tubes, bent aluminium has no competitor at present. However, diffusing reflectors of this kind are usually made of stove enamelled sheet steel and where possible the reflector is a mechanically strong part of the housing. Plastics reflectors are also used to some extent and at one time there was a vogue for long translucent plastics diffusing troughs, open bottomed. The trough provided a cut-off to the bare lamp and emitted a pleasant upward component of light to reduce the tunnel effect. This translucent design seems now to have gone from fashion however; slotted top opaque troughs are commonly used in industrial surroundings and prismatic or louvered units in commercial premises.

The properties of plastics and of glass are examined in more detail in chapters 18 and 19; in the next chapter we shall consider some of the mechanical and electrical requirements of modern design and examine testing methods for ensuring performance and durability.

7 Safety and Durability

The optical system and its light distribution is not the only matter of importance in designing a luminaire. For some purposes other requirements are paramount; for example, in some hazardous locations where flammable vapours are present, the problem is to construct a fitting which will not significantly increase the likelihood of explosion. Such conditions may occur in coal mines, oil refineries, paint spraying shops, at petrol pumps and so on.

Dangers in use

In some instances the great problem is rough handling of the lighting equipment—robustness rather than elegant light control may be the hallmark of equipment for cargo handling areas on ships or for the temporary lighting of building sites. Corrosion is another environmental hazard which has a major influence on design; the many different types of chemical works present their own special and different difficulties. Vibration, not only on trains, ships or motor vehicles but also in industrial buildings where it can sometimes be exceptionally severe, will also affect the requirements of design, although seldom such as to limit the optical control.

At the other end of the scale, as it were, we have the enormous range (and quantity) of decorative fittings in which precise optical control is not needed but where appearance is all-important. The diversity of types is impressive; modern shapes sometimes seem to have no connection at all with illumination. There is the inexhaustible domestic standby, the pendant fitting, with every variety of material formed into an infinite diversity of shades hanging from flexible cables. In such fittings (many of them really deserve the name "luminaire") the optics are almost negligible in their simplicity but the demand for variety and cheapness brings its hazards. The dangers of poor insulation and of overheating to the point of a fire risk are greater than most users realize. The risks of electric shock were always present but are increased nowadays by the greater use of central heating which provides an excellent earth connection. The human body may be an all-too-efficient circuit between a radiator and an accidentally live part on a lighting fitting. On the Continent these dangers have been recognized for some years and in the EEC strong controls are placed on the sale of equipment (especially domestic equipment) which must reach a suitable standard of quality. The greatest emphasis is placed on electrical safety and on fire risk— wooden houses are more common abroad than in Great Britain— but many other factors are now also controlled by specification.

International specifications

The main drive in this work came from the International Commission on Rules for the Approval of Electrical Equipment (abbreviated to CEE) which is composed of "organizations in European countries which, in the interest of the public and especially with regard to safety, issue rules and regulations for electrical equipment (cables and flexible cords, accessories and appliances) and check compliance therewith, as far as such checking takes place. CEE Specifications are mainly concerned with safety requirements and, although they are not formally binding on member organizations, the appropriate authorities in their countries are strongly recommended to adopt the CEE Specifications as far as practicable". The member organizations are for the most part the test houses. These are testing laboratories set up with governmental support and authority, whose task is to ensure that equipment passes the necessary tests. The United Kingdom participates in this scheme, our link with the other countries being the British Standards Institution and their laboratory at Hemel Hempstead.

In 1963, after several years of committee work, the CEE published a Specification for lighting fittings for incandescent lamps for domestic and similar purposes (publication 25). Two years later the International Electrotechnical Committee (IEC) issued its publication 162, Lighting fittings for tubular fluorescent lamps. The IEC has similar objectives to CEE but is more widely based geographically; it is also concerned with aspects of quality beyond safety. However, it is clearly an untidy arrangement for two bodies to issue international specifications on almost the same subject (luminaires are by no means the only overlap between the two bodies) and for several years attempts have been made to harmonize the two documents. A large measure of agreement has now been achieved.

British specifications

During this period it has become the accepted policy of the British Standards Institution to base its specifications as far as is practicable on international specifications. Hence the most recent standard BS 4533, Specification for electric luminaires (lighting fittings), conforms closely to what is widely accepted internationally. A résumé of some of its more important sections will give a useful indication of present thinking on luminaire design.

The British Standard, which is issued in loose-leaf form, will eventually cover the whole range of luminaires. Most of the information issued at present is concerned with the general requirements and tests in Part 1, but this is being followed in Part 2 by sections dealing with particular types.

The general introduction states that the standard is based on the requirements of safety, durability, performance and ease of maintenance. It goes on, however, to make the important point that no luminaire can be expected to remain in good condition without adequate maintenance.

After a section on definitions, the standard proceeds to classification under three headings: type of protection against electric shock; protection against ingress of moisture and dust; material of supporting surface.

Protection against electric shock

The standard defines and immediately excludes a form of luminaire very common in this country; this is Class O and its sub-class OI which has only functional insulation and no arrangement for earthing. Functional insulation is simply the "insulation necessary for the proper functioning of the luminaire and for basic protection against electric shock". If this insulation should fail, for example if the sheathing of the conductors breaks down, the metal parts may become live. The risks of this type of construction are perhaps not severe for ceiling mounted fittings unlikely to be touched but, for example, metal parts of portable lamps present a much more serious problem. Modern good practice has therefore swung right away from Class O luminaires and good wiring practice always provides an earth wire with the mains supply.

Class I equipment has functional insulation plus earthing and now becomes standard practice. Class II luminaires have functional insulation plus a supplementary insulation for extra safety—a form of belt and braces. Alternatively they may have reinforced insulation which both electrically and mechanically is as good as double insulation. Such equipment is not earthed.

Finally, the standard recognizes Class III low voltage luminaires either 50 V d.c. or 50 V a.c. with less than 30 V to earth.

Thus with Class I luminaires, if the metalwork becomes live on the normal 250 V supply (the maximum value with which this standard deals), a fuse will blow and the luminaire become dead. With Class II, the exposed metalwork (if any) is very unlikely to become live; with Class III, the voltage to earth is too low for live metalwork to present a hazard.

Marking

Symbols indicating the classification must be marked indelibly on the luminaire, together with a variety of other information such as trade mark, supply voltage and rated wattage. Other information which may be given includes supply frequency, operating temperature, maximum safe ambient temperature, and any special conditions of use; specific tests are detailed to ensure that the markings are indelible.

Electrical and mechanical design

There is a whole series of requirements to be checked by inspection or test of which the following are some examples: wireways to be smooth and free from sharp edges; switches to be securely fixed and adequately rated; insulated linings to be securely fixed and strong enough; live parts not to be in direct contact with wood; materials which burn fiercely such as celluloid not to be used for shades. Screwed connections must withstand the stresses of normal use (checked by tightening and loosening the screw using a screwdriver applying a specific torque); the length of engagement of a screw in a thread of insulating material must be at least 3 mm plus one-third of the screw diameter; thread-cutting screws may not be used to connect current-carrying parts.

Mechanical strength is proved by blows from a spring-operated hammer whose construction and method of use is carefully detailed and, in addition, a torque test is applied to screwed glands. A suspended luminaire must be capable of supporting four times its own weight for one hour and rigid suspensions are subjected to other loading and torque tests. A luminaire suspended only by a flexible conductor may not exceed 5 kg weight and adjustable fittings must withstand 10 000 complete operations from one end of the range of adjustment to the other. Apart from all these requirements, an Appendix gives a guide to good practice in luminaire construction, advising on choice of materials and methods of protection against corrosion. However well a specification is devised, much depends on the skill and honesty of manufacture.

Later sections continue with various provisions regarding electrical safety. Cable sizes are quoted in detail and tests given to ensure that cables are anchored firmly where this is needed. Each cord is pulled 100 times for 1 second each time at a tension up to 120 newtons (about 27 lb). Tests are applied to ensure that supply connections and terminals do their job properly, that the earthing is good enough and that live parts are effectively screened against accidental touching—this last requirement is checked by a special test finger.

There is a section giving a variety of tests to ensure that the luminaire lives up to its claims to be drip proof, rain proof, splash proof, water tight or jet proof, involving artificial rain in various forms, hosing or total immersion. To check for the dust proof or dust tight quality, the luminaires are put into a talcum powder dust chamber and connected to a vacuum pump to draw air through. None of these tests must leave the unit significantly the worse for wear.

Finally all luminaires must go for 2 to 7 days into a cabinet with a relative humidity of 91 per cent to 93 per cent between 20°C and 30°C, without showing damage, and subsequently they must withstand the special tests designed to check the insulation resistance. Measurements of resistance are made between the various parts of the luminaire and voltages up to 4·5 kV applied. Leakage current is also measured. Then comes the last of the purely electrical checks, namely, the measurement of creepage and clearance distances between live parts, to limit the possibility of tracking from one metal part to another. There are more than two dozen electrical tests in all.

Heat dissipation

A special section deals with the very important subject of operating temperature. Internationally this is a matter which has given rise to long and difficult discussion because in countries using much wood in their buildings—such as in Scandinavia—the attitude is naturally even more cautious than in the UK. Such matters as what overload in supply voltage is to be assumed; what are the safe operating temperatures for p.v.c., acrylic plastics, paper and fabrics; what hazardous failures of part of the equipment are likely (e.g. failure of a fluorescent lamp cathode); what ambient temperatures are to be regarded as normal; all these lead to a great variety of points of view. It is by no means easy to resolve the difficulties and to find a good compromise between safety and the many other factors involved, such as cost and convenience.

Temperature tests are made with the luminaire in a double walled enclosure which permits free circulation of air without draughts, using thermocouples of specified type and size. A full list of limiting temperatures is given, within which the various parts of the luminaire must operate. For example, a cemented lamp cap must not be hotter than 210°C; p.v.c. insulation is limited to 90°C or to 70°C if subjected to mechanical stress; acrylic plastics may not operate above 90°C or polystyrene above 75°C and so on.

Ballasts and capacitors must operate below the safe temperature claimed by the manufacturer and marked on the case (they undergo tests of their own to establish safe operating temperatures). If there is no such marking, ballasts are limited to 85°C or 95°C according to their construction, and capacitors to 50°C. Following the temperature measurements, an overload voltage of 10 per cent is applied for five periods of 24 hours each at the end of which there must be no important deterioration in the luminaire and it must also withstand a high voltage test.

It might seem that by now the luminaire had been tortured sufficiently, but it must still satisfy the next section—resistance to heat, fire and tracking. All external parts of insulating material protecting against electric shock, and all other insulating parts which retain live parts in position, are subject to a ball pressure test. A 5 mm steel ball is pressed into them with a force of 20 N for one hour at 20°C above normal operating temperature; the dent must not exceed 2 mm diameter.

For resistance to tracking, two platinum electrodes 4 mm apart are pressed on the surface at room temperature and an ammonium chloride solution is dripped between them so that 50 drops fall in about 25 minutes; no flashover must occur with 175 V a.c.

By comparison, the resistance to fire of the insulating parts is checked quite simply; they must be self-extinguishing.

Standards for particular fittings

Part 2 of BS 4533 includes the sections on particular types of luminaire and six of these sections are already issued: (2.1) Lighting fittings for Division 2 areas, (2.2) General purpose luminaires, (2.3) Electric handlamps, (2.4) Portable luminaires, (2.5) Floodlights, (2.6) Luminaire track systems. These sections in Part 2 call up the appropriate tests from the general requirements in Part 1 and also include additional tests or modify those in Part 1 as appropriate.

It will be seen that a luminaire must now run the gauntlet of a very full and severe examination. In some ways this is a far cry from the old days when the reputation of the lighting fitting manufacturer was enough for most customers. This attitude was justified because the technical morality of British manufacturers was high and they could be trusted to produce safe and reliable equipment for the most part. But the wider European and indeed the world markets have forced us to adopt the same detailed standards which the Continent deems necessary in order to suppress inferior products.

The expertise of British industry is such that, although it may have to change its thinking to some extent, it will not find grave difficulty in conforming to the new requirements. But if it were not to conform, it would find itself debarred from many overseas markets.

It is to be hoped that users and legislators will recognize the need to apply these standards in the UK also, and will not be tempted by the cheaper and nastier, whether made at home or abroad, in order to save a few pence. Penny wise, pound foolish is still true, even in metric units.

8 Lamps

Lamps create light, by consuming electricity or other forms of energy; they convert electrical power or thermal power into luminous power, which we call luminous flux and which we measure in lumens. They produce the raw material of illuminating engineering, they are the practical basis of lighting and the foundation on which most of our techniques of light control and distribution are built. Even though daylight is said to be "free", we need lamps in daytime as well as at night. Most of the major advances in lighting practice have occurred because the development of new lamps made them possible.

The enormous growth of lighting, which has itself made much of our 20th century industrial life possible, has been due largely to the enormous increase in the efficiency of lamps (i.e. the conversion rate from electrical energy to luminous energy). The figures are impressive:

 1880 2 lumens per watt (lm/W)
 1900 4 lm/W
 1920 12 lm/W
 1940 40 lm/W (or 80 lm/W for sodium yellow light)

1960 60 lm/W (or 100 lm/W)
1970 90 lm/W (or 150 lm/W)

Even though lamps and electric power have become cheaper in this period, this is the item that shows the biggest increase in value. Furthermore, our 1000 lux installations would be unbearably hot, as well as too expensive, if we had to use the lamps of even 40 years ago.

More generally, lamps create radiation of which part is perceived as light and part is perceived only as heat, and the lamps themselves get hot in the process. The higher the proportion of visible radiation, the greater the efficacy, the more lumens per watt and the cheaper the light (unless of course the more efficient lamp is more expensive).

Sources of energy

Hot bodies emit radiation whether they are heated mechanically as in a flint-and-steel spark, or chemically as in a flame, or electrically as in a filament; their light output from a given area depends mostly on their temperature. There are certain materials such as cerium, which is used in gas mantles, which emit relatively less of the unwanted infrared than other materials such as carbon or alumina, and therefore the mantle gets hotter and emits more light. Tungsten is fortunately one such material, emitting approximately only half as much infrared as a theoretical "black body" emitter of the same light output, and therefore requiring less power to operate at a given true temperature.

Light derived from incandescence has a continuous spectrum the energy distribution of which depends mostly on the temperature of the hot source, so that the colour quality of the light can be described fully by a single figure on the Kelvin temperature scale—the colour temperature.

The other principal source of light is the excitation of electrons in the atom, such as that produced by an electric discharge passing through a gas or through a metallic vapour. When an electric potential is applied across a discharge lamp, the electrons which are emitted by the electrodes are attracted and accelerated through the gas to whichever electrode is positive at that moment; they collide with the atoms of the gas and excite them to a higher energy state. The same occurs with atoms of the metallic vapour in a discharge lamp. When the atoms return to their original energy state, they emit their excess energy as a pulse of radiation which may have a wavelength in the visible region of the spectrum (high pressure mercury lamps) or in the ultraviolet (low pressure mercury as in fluorescent tubes).

Similarly, the absorption of short wave radiation (ultraviolet) by a crystal of a fluorescent material alters the energy states between the atoms of the crystal and light is radiated as they return to normal, either immediately (fluorescence) or over a period of time (phosphorescence). This is the process which produces the light in fluorescent tubular lamps; the electric current generates ultraviolet radiation in the low pressure mercury vapour and this radiation causes fluorescence of the powder layer inside the tube.

This light does not have a continuous spectrum similar to that from a hot body but consists of either a line spectrum emitted at specific wavelengths or a band spectrum extending over restricted ranges of wavelengths. The colour quality of the light depends both on the activated material itself and on the physical conditions of activation. By the right choice, a major part of the re-radiated or emitted energy can be confined to the visible region of the spectrum, giving a good energy conversion rate and therefore a high efficacy, although there are natural laws which prevent 100 per cent conversion. The control of the energy distribution in the visible region (that is of the colour appearance and the colour rendering of discharge and fluorescent lamps) and in addition the control of efficacy and long term stability are unavoidably complex in manufacture and in use.

Output of energy

Conservation of energy means that every packet of energy that goes into a lamp—calorie or joule or kW h—must come out again in some form, either as radiation through the surrounding air (ultraviolet and infrared as well as visible) or as sensible heat conducted through the lamp holder or convected away in the surrounding air. Some of the radiation created inside the lamp may be converted into heat before it gets out, such as the short wave ultraviolet or the long wave infrared which are absorbed by the glass envelope and either emitted as very long wave radiation or convected away as sensible heat in the air.

The cold lamp in which all the input energy is converted into light, which in turn is freely transmitted through the bulb without any absorption that would even make it warm, is still a long way off and any approach to it will doubtless involve different techniques from conventional lamps. If all the energy were emitted as green light, the efficacy might be as high as 680 lm/W, the theoretical maximum which is one of the fundamental figures of

Table 8.1 Output power from lamps in free air as a percentage of input wattage

Type of lamp	Radiated power			Conducted and convected power
	ultraviolet	visible	infrared	
Incandescent				
GLS and TH	0·5	6	74	20
Mercury discharge				
MA and MB	10	10	50	30
MBF	1	15	54	30
MBIF	1	20	49	30
MCF	2	23	30	45
Sodium discharge				
HPS or SON	—	22	40	38
SOX	—	30	30	40

photometry (the absolute value of this maximum is known to better than ± 1 per cent). An idealized sodium lamp with no heat production whatsoever would give rather more than three-quarters of this, namely 520 lm/W. If the energy were all within the visible region of the spectrum and so distributed as to be white, the theoretical maximum efficacy is between 200 and 250 lm/W, depending on what is meant by "white". The lamp technologists have developed conventional lamps up to about one-third of the ultimate and unattainable ideal, which is a notable achievement.

Table 8.1 indicates very broadly the distribution of output power from lamps in free air, as percentages of the input wattage.

Incandescence and tungsten lamps

Red-hot and white-hot are self-explanatory, although we have the strange anomaly that the higher colour temperatures are regarded as the "cooler" light ("The cold light of day"). The proportions of infrared, of the various wavelengths in the visible, and of ultra-violet are determinable from Planck's law with great precision.

The commonest lamp, the tungsten filament lamp, has reached a very high level of sophisticated manufacture. Tungsten is a semi-ductile metal which melts at a high temperature (3650 K); its electrical conductivity decreases as it gets hotter, so that if a constant voltage is applied the input current decreases as the temperature rises until there is equilibrium between the input energy and the radiated energy. This equilibrium is very stable so that consistent manufacture is possible. The crystalline structure of the filament keeps it rigid up to near its melting point, although vibration can cause the crystals to slip which leads to filament sag or to local hot spots; this can be counteracted by traces of other metals which control the crystal growth, and also of course by proper mechanical supports. The metal evaporates somewhat before it melts, leading to blackening of the bulb, but the rate of evaporation can be reduced by surrounding the filament by an inert gas (nitrogen or argon); the filament is cooled by convection so it has to be coiled or double coiled to reduce the transfer of heat energy to the gas. The filling gas must of course be very pure, in particular anything containing oxygen must be wholly excluded if the maximum life is to be attained. The glass bulb contains the gas, protects the filament and its supporting structure, sometimes diffuses the light, and allows us to handle the whole lamp. It all seems so simple but in practice there are about 4000 different designs of tungsten filament lamp in manufacturers' lists and of course in demand, often irregularly, by the users.

The effects of evaporation can be counteracted by putting a small amount of a halogen vapour (bromine or iodine) in the bulb; the tungsten vapour combines with the halogen instead of condensing on the inside of the glass bulb so that blackening does not occur. The tungsten–halogen compound dissociates when the internal convection currents in the gas filling carry it near the hot filament, creating a concentration of tungsten vapour around the filament and so reducing the rate of evaporation. This permits a higher temperature and therefore a higher efficacy for a given lamp life and, because the bulb stays clean, gives constant light output through life.

The reasons why a tungsten filament has a limited life are not all known but it does break after a period of time which depends on the design and on the operating temperature, amongst other factors. If it is too hot (over-run), its life is shorter but it gives more light; if it is under-run to give a longer life, its light output and its efficacy as a light source are reduced. The economic balance for 240 V GLS (General Lighting Service) lamps on a strict cost/benefit basis at 1973 prices was about 500 hours for the higher wattages and about 1300 hours for the lower wattages but the lamp manufacturers adopt a target of 1000 hours for the whole range. The compromise between life and light is clearly demonstrated by the ranges of tungsten halogen lamps, some of which have short and very merry lives whilst other are rated at 2000 or more hours because their economic balance is at a lower light output and efficacy.

GLS lamps fail sooner if operated upside down (i.e. cap down in the usual design) or on their side; this is due to the changed flow of internal convection currents and to the mechanical support of the filament being inappropriate to the abnormal position. The reduction of life on a test rack is about 10 to 15 per cent but it seems to be more in practice. Tungsten lamps also fail sooner if operated in a confined space, not so much due to the filament running hotter (by calculation, 20°C increase in ambient temperature only means about 0·2° increase in actual filament temperature or under 1 per cent decrease in life) but rather to the glass bulb being hotter, which affects the internal convection cooling and also the purity of the filling gas. So tungsten lamps burn best the right way up in cool surroundings; they work perfectly well in a refrigerator and they suffer no damage in a baker's oven unless they are switched on when the oven is hot or unless the oven is too hot for the lamp cap (about 160°C).

Glass bulbs get hot anyway, by convected heat from the gas filling but even more by absorption of the infrared and ultraviolet radiation from the filament. This does not have any adverse effect up to about 300°C, or to 400°C for heat resisting glass bulbs, but glass ceases to be an inert, stable, impervious material above temperatures of this order. Even the electrical insulation in the glass pinch which supports the filament and its supply wires can start to weaken at abnormally high temperatures. Fused silica bulbs as used for halogen lamps are stable up to a dull red heat.

Effects of voltage change

Incandescent tungsten lamps are sensitive to the supply voltage and a maintained increase of 1 per cent in the voltage has the following effects:

current	0·6 per cent increase
wattage	1·6 per cent increase
light output	3·7 per cent increase
life	14 per cent decrease
colour temperature	12° rise

The cold resistance of a filament lamp is very much less than its resistance at operating temperatures, by a factor of 0·10 to 0·07, so there is a surge of current for a fraction of a second when a lamp is first switched on. The duration of this surge only becomes a problem with heavy current lamps in which thick filaments take longer to heat up.

Light-controlling lamps

Pearl bulbs and white-coated bulbs provide diffusion which improves the appearance without seriously reducing the light output (less than 2 per cent for pearl and about 6 per cent for silica or titania powder coatings). A pearl GLS lamp emits approximately the same intensity in all directions, except that of the cap end, but a white bulb lamp or a long linear filament lamp emits its maximum intensity sideways.

Reflector bulb lamps have a film of aluminium deposited on part of the inside surface of the bulb, which may be shaped so that the desired optical control is obtained; either back reflection from a "crown silvered" lamp or a forward beam from a paraboloid bulb lamp. The spread of the beam depends on the size of the filament, the detail shape of the reflecting surface and the diffusion at the front surface of the bulb. The sealed beam lamp with its pressed glass bulb is particularly precise in its optical control, as well as having a desirable robustness.

Selective reflection may be obtained by the use of a "dichroic" film deposited on the inner surface of the bulb. This film comprises very thin alternate layers of two transparent materials of different refractive index, deposited by a process of evaporation under conditions of high vacuum. Each layer has a thickness equivalent to $0.25 \times$ the wavelength of the light to be reflected and by using multiple layers only a restricted range of wavelengths is reflected and the rest transmitted. Conversely, a restricted range may be transmitted and the rest reflected. With the right choice of thickness most of the visible light can be reflected whilst the infrared is transmitted so that a beam of cool light is projected. Such a lamp will of course emit more heat through its reflecting surface than an aluminized lamp and the housing may get unexpectedly hot. A similar technique can be employed to produce a coloured beam and, as there is no actual absorption of radiation in the multi-layer film, the lamp generally runs cooler and gives more light than a dye- or pigment-varnished lamp.

9 Discharge Lamps

The discharge is the oldest form of electric light, whether it be a flash of lightning, the Aurora Borealis or the glow of Watson's enclosed spark "lamp" of 1752. If we go further back, the Divine command "Let there be light" reduced chaos to ordered energy radiation from electrical discharges rather than from incandescence; the discharge lamp has a very respectable ancestry.

The Davy carbon arc of 1810, the Geissler discharge tube of 1860 (which Becquerel used for experiments on fluorescence) and the Birch open arc of 1876 were all earlier than the incandescent filament lamp invented by Swan and separately by Edison in 1878. In this century the early discharge lamps were long high voltage tubes such as the Moore white carbon dioxide lamp, made up to 180 ft long in 1904, or the Claude red neon and blue mercury–neon lamps of about 1910 which are still familiar to us. The Cooper-Hewitt low pressure mercury arc was the first production mains voltage discharge lamp at about the same time: mercury has remained a principal constituent of discharge lamps ever since, such as in the high pressure type MA of the 1930's and in the types MCF, MBF and MBI of the succeeding decades.

The other principal discharge lamp is the sodium vapour type, introduced early in the 1930's as a low pressure lamp of long life, high efficacy and monochromatic yellow colour and developed in the 1960's as a high pressure lamp of greatly improved colour. This is a notable history of achievement.

This chapter does not discuss the xenon discharge lamps which are very effective in certain specialized applications but are not in general lighting use. They have remarkable properties such as instant lighting, high luminance and excellent colour rendering but they are uneconomic for general application.

Electric discharge

Current can be conducted through a gas or vapour between two electrodes under the right circumstances; it so agitates and excites the gas that light is emitted. This is different from heating a filament to incandescence because the activity occurs within the atoms of the gas and causes each atom to radiate in its own way. The greater the current, the greater the activity and the better the conduction, so that unless there is some means of stabilizing the current, a constant voltage applied to a discharge lamp would cause it to fail catastrophically.

The current is carried largely by (negative) electrons moving through the gas as in a metal filament but with the difference that a source of electrons has to be available in the gas before the current can start; the current is also carried partly but importantly by positive ions, which are atoms of the gas with a positive charge due to the release of an electron. The conductivity depends critically on the production of ions and release of electrons by collision between the electrons and the atoms of the gas, so a delicate balance must be maintained between voltage and current, pressure and temperature, and of course the amount of the gas or vapour present.

There is no need for confusion between a gas and a metallic vapour; both are gaseous and act like gases when hot. When the lamp is cold the gas is still there but the vapour has partly condensed to the metallic state. A gas such as argon or neon is necessary to start the discharge when cold, after which the metal evaporates as the lamp warms up and the current is carried through the vapour, not of course through the droplets or particles of the metal. It is the collisions between the electrons and the atoms of the vapour which set up changes in the energy states of the atoms and cause them to emit light.

The varieties of electrodes, of gases or vapours and of operating conditions can be extremely wide; the theories of glow discharge, arcs, positive columns and cathode falls are too complex to be examined here. The commercial pressures for high efficacy and for whitish light have reduced the variety to the vapours of mercury or sodium with suitable proportions of other gases or metallic vapours which either facilitate the electrical operation or improve the light emission. Mercury is a principal constituent in these practical discharge lamps (there is mercury as well as sodium in a high pressure sodium lamp) and it may be activated in three ways. At very low pressure, it emits ultraviolet radiation; at moderate pressures, it emits the familiar yellow, green and violet lines of the visible mercury spectrum (perhaps not so familiar because these lamps are not in current use) and at high pressure it emits bands of radiation throughout the visible spectrum but still with emphasis in the green and blue rather than in the orange and red. At low pressures the lamp can start almost immediately because there is enough vapour pressure at room temperature for the discharge to operate, but high pressure lamps take more time; they may require several minutes for the temperature to rise so that the proper amount of mercury evaporates. When a high pressure lamp is switched off, the presence of the vapour and the absence of free electrons or ionized atoms prevents the passage of current-carrying electrons until the temperature and pressure have fallen, although a high voltage impulse may be used to restart a hot lamp. High pressure lamps necessarily run hot and are not greatly affected by the ambient temperature, although they may not start below $-30°C$. Low pressure lamps are very sensitive to temperature; if they run cool, the vapour pressure is inadequate and the current and wattage are greatly reduced, and if they run hot the excessive vapour pressure inhibits the proper emission of ultraviolet light.

The vapour is necessarily enclosed in a transparent tube with electrodes at the ends; a glass tube is used for low pressure lamps and a fused silica or alumina tube for high pressure, hot lamps. High pressure lamps have an additional outer bulb which serves to protect their relatively delicate construction, to conserve heat and to absorb any troublesome ultraviolet radiation.

At any particular temperature and pressure all practical discharge lamps have a negative resistance characteristic (i.e. conductivity increases as voltage rises) and therefore some current-limiting ballast is essential. This may be a resistance, as in the MBT series of lamps which have a tungsten filament in series with the discharge or as in some of the less expensive fluorescent lamp fittings. A much more economical and often more convenient device is an inductive choke or auto-transformer, usually combined with a capacitor, which has a much lower power loss and may give advantageous starting characteristics. The complications of the ionic discharge and the inductive control can lead to great distortions in wave form, sometimes with a substantial third harmonic component, and to low power factors, so there is scope for great ingenuity in the design of the control gear.

Having overcome all these difficulties, discharge lamps are in many ways more robust and enduring than the much simpler incandescent tungsten filament lamps, in addition to having greater luminous efficacy.

Luminescence and fluorescence

Certain crystals can produce light if activated by electrical forces (electro-luminescence) or by ultraviolet light (fluorescence and phosphorescence). The low pressure mercury discharge in tubular fluorescent lamps converts about half its electrical energy into ultraviolet light which is then completely absorbed by the layer of crystals coated on the inside of the tube; this layer fluoresces and is the source of light.

Again we have a complicated balance between the activating radiation and the chemical composition of the crystals, including trace elements as activators, together with the crystalline state and the operating temperature. The efficacy and lumen maintenance of the lamp are determined largely by the fluorescent powder, although lamp design is a factor affecting these characteristics. The choice of the crystalline material must be a compromise between the overall amount of light produced and the constitution of that light, because there would be little benefit in producing a large amount of unacceptable light, that is light of the wrong colour properties.

With high pressure mercury lamps the fluorescent powder on the inside of the outer bulb increases the total amount of light produced, improves the colour by adding orange and red to the light from the mercury discharge itself and reduces the flicker by its phosphorescent effect. These benefits have also been gained in the mercury halide lamps and in the tungsten filament ballasted lamps, where the ultraviolet emission from the discharge can be converted into light rather than just being absorbed by the outer bulb.

Fluorescent light is in some ways more under the control of the lamp designer than either incandescent or discharge light. The emission is in broad bands of the spectrum rather than in spectrum lines so a continuous spectrum can be built up from several fluorescent compounds to give a light which is rich in the red or green or blue regions and which is much more acceptable to the human eye than monochromatic light or a discontinuous spectrum would be. This great advantage has led to the production of a superabundance of tubular fluorescent types, of different colour appearances and of different colour rendering properties, which may not be easily distinguishable or identifiable to many of us. The eye is very tolerant of these differences and we can see equally well in a mixture of "warm white" and of "daylight" lamps, or we may not even notice the mixture of good and bad colour rendering lamps. The mind is, however, less tolerant and we may be irritated or confused by carelessness in choosing, or failing to choose, the right kind of lamp. Further, some lamps may be markedly deficient in output in some restricted region of the spectrum without this being fully appreciated, such as the lack of extreme red from some high pressure types of lamp, which may give a distorted form of colour rendering or may interfere with the subconscious muscular reactions which keep our eyes in focus; these are matters which are of increasing importance as electric lighting comes to play a bigger part in our lives, and the manufacturers of the lamps are treating them seriously. The problem is to improve on what many regard as the ideal light—daylight—at the same time as providing electric lighting installations which suit us better than natural lighting would. Fluorescent light is capable of blending very well with daylight, both sunshine and north sky, but it is not to be regarded as the same as daylight; in some ways it is worse but in many ways it may be better.

Temperature effects

All lamps get warm or hot and the luminaires around them get hot too; this rise in ambient temperature around the lamp may make an appreciable change in its performance.

We have seen in Chapter 8 that the light output from incandescent tungsten lamps is not significantly altered by ordinary changes in ambient temperature although their life may be reduced. On the other hand, the life of discharge lamps is not generally affected by ordinary hot environments, within the limits set by the glass to metal seals or by the lamp caps, but their performance may be affected considerably. Most serious is the effect on tubular fluorescent lamps which give only half their standard output if the ambient temperature around the tube is either 25°C below or 40°C above their preferred ambient of 25–30°C; this may give a very misleading distortion of the measured light output ratio of a luminaire. The upper limit may be raised by using amalgam lamps, which contain an indium mercury amalgam whose vapour pressure is less than that of pure mercury and for which the preferred or optimum ambient is 45–55°C, or by cooling the tube locally by a heat sink or a well-ventilated spot which will reduce the mercury vapour pressure throughout the whole tube.

High pressure mercury lamps are not significantly affected by ambient temperature. Both low pressure and high pressure sodium lamps give more light when in a warm environment, over 5 per cent in some instances, which can also distort the light output ratio of a street lighting lantern but in a more favourable sense.

Ballasts increase in electrical resistance and in power loss when hot, which reduces the output from the lamp and shortens the life of the ballast. They are also incidentally a potential fire hazard in that some kinds of lamp failure or ballast failure cause overheating of the ballast rather than of the lamp; proper provision has to be made for the dissipation of heat from the ballast, particularly in luminaires to be mounted on heat-sensitive surfaces. It is in the study of the performance of lamps and their control gear that the thermal problems of luminaire design become most apparent.

Effects of voltage change

Discharge lamps are less affected by changes in the supply voltage than incandescent lamps and it is not unknown for them to be overrun, with consequent damage to their control gear. A maintained increase of 1 per cent in the supply voltage increases the light output by about 0·5 per cent for low pressure sodium, 1 per cent for tubular fluorescent, 2 per cent for high pressure mercury and 3 per cent for high pressure sodium and tungsten filament ballasted high pressure mercury, type MBFT. The lamp life is very slightly reduced except for the type MBFT lamp, the life of which would be reduced by about 10 per cent.

Increased supply voltage also increases the lamp current, more so than with incandescent lamps; 1 per cent additional voltage gives 1·5 to 2 per cent increase in the current and two or three times this in the power loss in the ballast, but these figures are only representative of the general trend and may not apply to a particular case.

In an installation of tubular fluorescent lamps in enclosed luminaires the increase in light output due to an increase in supply voltage may disappear after a short time as the additional circuit wattage causes the operating temperature to rise and so reduces the light output and the efficacy of the lamps; again, the thermal problems of the whole installation have to be considered.

Lamp performance

This chapter has not attempted to give performance data for particular lamp types but only general information on which the proper detailed figures may be appraised. The variety is so wide that the manufacturers' catalogues are the best sources of information. The several British Standards deal with dimensions, wattages and interchangeability but they set minimum values of light output rather than average or target values. Most discharge lamps have long lives to burn-out but their light output falls to uneconomically low values long before they fail. The usual basis for stating life duration is a fall to a given percentage of the initial light output rather than the period in which a given percentage of lamps fail completely, but there is no widely established figure for the "given percentage fall in light output". The usual basis for stating the nominal light output of a type of lamp is no longer the "average through life" but is a notional figure called the "lighting design lumens" which is generally equivalent to the output after 2000 hours burning and is representative of the average to be expected during the nominal life of the lamp. This is the figure to be used in calculating the service illuminance as recommended in the IES Code.

The British Standard specifications are designed to control the quality of lamps through some rather elaborate statistical techniques in which the values stated for life or lumen maintenance or light output are not always applicable to the practical conditions of use; guidance on performance figures may be obtained from the Lighting Industry Federation or from individual manufacturers.

10 Scalar and Vector

The 1973 IES Code recommends "scalar illuminance" as a measure of the lighting of some interiors and also refers to "illumination vector" when discussing the quality of lighting.

These are new words for what are really quite old* and familiar ideas, namely the lighting of solid objects and the flow of light. But the words "scalar" and "vector" are unfamiliar and to combine one of them with "illuminance" and the other with "illumination" is to make matters worse for new readers of the Code.

Illuminance and illumination

Let us deal with these nouns first before we start on the adjectives. "Illumination" is the process of lighting, the application of light to an object so that it can be seen. "Illuminance" is the amount of light involved in the process, the number of lumens of light falling on unit area of the object, in lumens per square metre or lux. This amount used to be called "illumination" also but that was confusing, as H. C. Weston pointed out over 20 years ago. When those

---
* L. Weber in 1885 and A. A. Gershun in 1936.

innovators of the English language in the USA started to call it "illuminance", the rest of the English-speaking world found it convenient to follow.

Scalar illuminance

The dictionary helps: a scalar number is a real number, without dimensions or direction. In lighting, scalar means magnitude without direction, as in the temperature scale or values on the noise scale, so that the scalar illuminance at a point in space means the amount of light at that point irrespective of where it has come from. Formally, scalar illuminance is the average illuminance over the surface of a small sphere, but more of that later. It is also known as "mean spherical illuminance", which is an obvious but rather clumsy term, unlikely to be used even in conversation among experts. Remember that scalar illuminance is nothing to do with the degree of diffusion in the light; it does not imply that the lighting is diffused and it does not relate only to the diffused component in the light. It is just a measure of the amount of light, irrespective of direction.

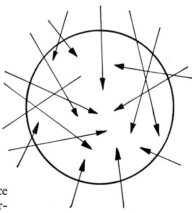

Fig. 10.1 Scalar illuminance (Flux through sphere/Area of surface)

Illumination vector

Again to the dictionary for the word "vector": a quantity having direction as well as magnitude. The flow of water in a river or the force exerted by a spring may be visualized easily, but the flow of light is more difficult. Light flows in so many directions, all at once, without mutual interference, but there is always (or nearly always) a predominant direction, a direction in which the light is flowing more strongly than in any other. This is the direction of the "illumination vector".

The magnitude of the "illumination vector" is quite simple; it is the difference in illuminance on the two sides of a plane surface which is facing up (and down) the direction of the vector.

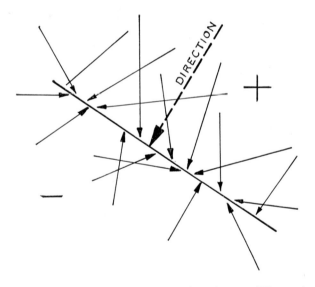

Fig. 10.2   Illumination vector (Maximum difference)

Planar illuminance

This term just had to come, by analogy with scalar illuminance, but there is nothing new in it. It is our old-fashioned illuminance, falling on a horizontal plane or a vertical plane or any other plane that we like to choose.

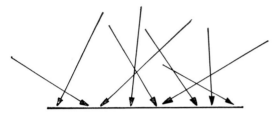

Fig. 10.3 Planar horizontal illuminance
(Flux on one side/Area)

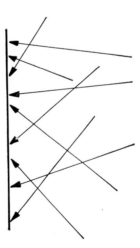

Fig. 10.4 Planar vertical illuminance
(Flux on one side/Area)

Cylindrical illuminance and conical illuminance

These terms haven't arrived yet, but they may, so we will define them in advance as the average illuminance over the surface of a small cylinder or cone respectively. Conventionally, the axis may be vertical and the apex angle of the cone may be $2 \times 15° = 30°$ but we need not worry about this until we have to.

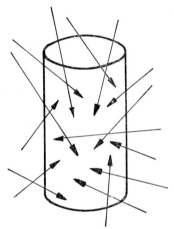

Fig. 10.5 Mean cylindrical illuminance (Flux through side/Surface area of side)

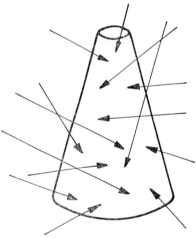

Fig. 10.6 Mean conical illuminance (Flux through cone/Surface area of cone)

Why do we need these ideas ?

The words "scalar" and "vector" came into our thinking in 1966 when J. A. Lynes and his colleagues W. Burt, C. Cuttle and G. K. Jackson of the Pilkington Brothers' Environmental Advisory Service read a paper to the IES showing how valuable these ideas are in the understanding of daylighting in interiors. This led of course to artificial lighting also. As an example, let us see how they come into daylight design.

Flow of daylight

Under an overcast sky, daylight flows principally downwards, just as the artificial light from a large installation of ceiling-mounted luminaires flows principally downwards, and it is reasonable to adopt the illuminance on a horizontal plane (E_H) to measure the amount of outdoor daylight. Indoors, however, the daylight from vertical windows flows obliquely and sometimes almost horizontally across a room. If we regard the window as the light source, we can understand that the amount of daylight available some distance from the window is not realistically measured by the illuminance on a horizontal surface. Lighting engineers often use illuminance on vertical surfaces (E_V); we know that a "well lit" room with "good windows" means that there is plenty of light on the far wall. Because vertical surfaces can face in different directions, the mean vertical illuminance or the average illuminance on a vertical cylinder has been used as a measure of the transverse flowing light.

However, sharp differentiation into horizontal and vertical surfaces is not true to life; most surfaces are oblique. The amount of light available at a point in space can be most realistically measured by the amount of flux passing through a small volume of the space at that point or, because we are mainly interested in the lighting of surfaces, by the amount of flux incident on the surface surrounding the small volume. Conveniently we assume this small volume in our space to be spherical in shape and we can then define the mean spherical illuminance or Scalar Illuminance (E_s) as the quotient of the amount of light in lumens passing through a small spherical volume by the area of the surface bounding that volume, in lumens per square metre or lux.

Experiments have established that the mean cylindrical or the mean spherical illuminance is a better measure of the apparent amount of general multi-purpose lighting available in a space than the horizontal illuminance. Each has its merits but the mean spherical has the additional merit of being entirely non-directional and therefore easier to calculate. This is why Lynes picked it out and called it "scalar" illuminance.

The particularly valuable property of daylight in a building is, however, its directional characteristic—its ability to reveal texture and shape so naturally. It is relatively easy to find the direction of flow at any point, but what is the amount? This property cannot be measured in terms of the illuminance on any single surface, whether flat, cylindrical or spherical. The flow of light can, however, be measured by the difference in the amounts on the front and back of a plane surface. If this surface is turned to produce the maximum difference, the normal to the surface will be the predominant direction of flow. The amount of the difference will be the strength of the directional component and will be the magnitude of the "illumination vector". It is not called "illuminance", because it is only the difference between two illuminances; the magnitude and direction taken together are a property of the illumination. Its direction is often more important than its magnitude.

These two properties, scalar illuminance and illumination vector, came from the need to express in numbers the more important quantities of daylight in buildings. The same is surely true of electric light, because designers of artificial lighting always have much to learn from a study of natural lighting, but the flow of electric light has not yet been examined with as much care as that of daylight.

Flow of electric light

When the objects to be lit are solid and directionless, such as vehicles in a car park, antiques in a shop or people in a hotel foyer, the scalar illuminance is often a better measure of the amount of light than the horizontal (planar) illuminance, which is why the IES Code recommendations for some situations are in scalar terms.

In any installation with fixed lighting positions, there is a fixed relationship between scalar and planar illuminances at any point, though this relationship may be difficult to measure accurately and almost impossible to calculate exactly. This is why we have managed fairly well in the past to light our hotel foyers, etc., in terms of the horizontal illuminance, but we all know that the aesthetics of lighting are not measured in this way: perhaps this is why many lighting designers distrust light meters. When we have had more experience with scalar light meters, we may find their readings are more informative.

When the objects to be lit require good modelling from the lighting, there must be a significant amount of directional light and the angular direction is important. The illumination vector can be used to express this in simple numbers, though there are some snags in this and we are not yet able to interpret the numbers as precisely as we should. Fortunately, the illumination vector is a true vector in the mathematical sense and it has the convenience that several illumination vectors at a point from several light sources can be combined by standard vectorial calculation. In this respect, it is easy to calculate. Our ability to interpret the calculations and measurements will come in time and the illumination vector will be a valuable tool in the hands of the lighting designer.

Even an elementary study of the scalar and vector magnitudes in a lighting installation can produce new understanding. For example, magnitudes in many modern offices are disconcertingly similar to those under a fully overcast sky. The significance of this merits further thought. To take a more extreme case, a floodlight in a sports stadium is strongly directional and we may wonder whether the techniques of good daylighting can be used as a guide to the design of sports lighting. We have the same problems of harsh or confusing shadows, of the proper illumination of near-vertical surfaces and of the many different directions of view, as well as danger of glare from a near-horizontal flow of light. The design of sports lighting is no less complex than the design of natural lighting in an office and similar criteria are involved. Just as a well-designed interior installation can be better than an overcast sky, perhaps a well-designed sports installation should be better too.

The 1973 IES Code anticipates knowledge that will grow during the next few years and introduces scalar and vector concepts in the expectation that we shall learn to understand and use them. In a sense, this is similar to the introduction of the IES Glare Index system in the 1961 Code, before most of us understood it or realized its full implications. Much importance was attached to the numerical values of the Glare Index in the 1960's and, as a result, remarkable developments were demanded and accepted in the optical design of luminaires, such as the prismatic "diffuser" or the parabolic metallic louvre. When we can measure quality, we will want quality as well as quantity.

If our use of scalar and vector quantities is going to improve our lighting systems, we shall first have to use them in full-scale practical installations. Theoretical advances get us nowhere unless we make practical advances—and progress can only be assured if we go to the trouble to assess what benefit we have gained.

Vector/scalar ratio

The effect of the illumination vector depends on its magnitude in relation to the total amount of light—i.e. in relation to the scalar illuminance. The vector is not itself a real illuminance but rather the difference between two illuminances—although its magnitude may be expressed in lux. It is the ratio of the vector magnitude to the scalar magnitude that indicates what proportion of the total light has a directional effect, and this is what the lighting designer wants to know. The vector/scalar ratio has magnitude and direction, like a vector, although the vector/scalar ratios from several sources are not additive.

The vector/scalar ratio is likely, therefore, to become important in electric lighting as a measure of its modelling potential and as a guide to the general density of shadows, but this is something that still requires a lot more study. Recent papers in *Lighting Research and Technology* show that our scientists are working on it and on other derived figures, such as the "effective vector/scalar ratio". What is really needed is full-scale experience that can be obtained from controlled experiments by lighting designers, supported by those who pay for their work.

The theory is sound, the practice is encouraging and these ideas will undoubtedly contribute to future improvements. But they are not yet fully understood by our scientists—for example, there have been no widely-published studies of the effects of inter-reflection on scalar illuminance. What is more important at this moment is that these rather academic ideas should be more widely understood, so that lighting technologists and lighting technicians can appraise the "amount" and "flow" of light in a more quantitative way than in the past. Again, a new issue of the IES Code is leading the way.

11 Daylight Design

Daylight, like any other natural resource, can be wasted by misuse. It is inexhaustible but it cannot be conserved or stored in the normal sense. Like other natural resources it is free, but there are unavoidable costs involved in getting it into our buildings and using it properly, both in first cost and in maintenance. Further, the windows that admit "free" daylight may also lose expensive heat during winter or admit excessive heat in summer. So design for daylighting is a proper subject for study and is an essential part of design for living in buildings.

The Romans knew that they needed natural light and knew how to space their buildings so that daylight could enter them and the spaces around them. In the ruined Temple of Apollo at Pompeii there is a tablet recording that a sum of 3000 sesterces was paid as compensation for infringement of natural light as a result of a wall being built around the temple by the city authorities. The problem is still with us, even though artificial light is now more readily available. Many reasons have been offered as justification for admitting natural light into buildings, including the promotion of good health, a feeling of deprivation in its absence and of course the belief that daylight is free.

Although the value of natural light has been appreciated for centuries, very little was done to predict its behaviour or how it would enter buildings until the latter part of the 19th century. The practical significance of the *ratio* of internal to external illuminance was realized early, as recorded by L. Weber in 1885, and was expressed in rule-of-thumb calculations of window size. In 1895, A. P. Trotter, Electrical Adviser to the Board of Trade (and a founder member of the IES in 1909), devised a portable photometer for measuring light intensities and suggested that it would be useful to know the daylight coefficients of certain buildings; these coefficients were the forerunners of our present daylight factors. The external and internal readings were to be measured simultaneously (Trotter obviously realized the variability of natural light); external readings were taken on the unobstructed roofs of the buildings or on the window sills, and internal readings were taken at selected points at table level (the controversial horizontal working plane). The measurements became known as roof ratios and sill ratios for obvious reasons—they were quoted as *ratios* because very wide variations in the absolute values of outdoor illuminance occur from season to season and hour by hour. It was also found that the photometric ratio was a better measure of the apparent amount of light inside a building than an absolute measurement. The roof ratio was very soon related to the natural illuminance received at some reference point in a building on a normal day during winter, which was the forerunner of our CIE Standard Overcast Sky and our nominal 5000 lux outdoor illuminance.

Interest in the lighting conditions within buildings grew, together with a desire to predict the performance under a variety of conditions, giving rise to many methods of predicting and of measuring daylight quantities. Following Trotter, we have men such as P. J. Waldram and J. Swarbrick in the architectural profession, the remarkable group of workers led by Allen, Beckett, Dufton, Hopkinson and Petherbridge at the Building Research Station, and many others across the world; all concerned with refined definition and measurement or prediction of both quantity and quality of the natural lighting conditions in existing or in proposed buildings.

Calculation of amount of daylight

Enough of history for the moment, but any lighting engineer trained only in electric light can learn a lot from those who have studied natural lighting. The IES library is a good starting point. Many of the traditions of architectural design and of pictorial composition have arisen from the flow of light through windows; the precision of daylight calculations is still as high as the general run of calculations for electric light. We have to separate direct sunshine from the diffused light from the sky, because the sun is subject to both complex movements and variations, and this helps to simplify the calculations and definitions.

Uniform Sky An overcast sky such that the sunlight and the blue sky are so fully diffused by clouds that it may be assumed to have the same luminance at every point. This never happens in real life and it is no longer used for accurate design purposes.

CIE Standard Overcast Sky The basis for design of daylighting in most temperate areas of the world is a "standard" sky, still with the sunlight obscured by cloud, which is only one-third as bright around the horizon as at the zenith. The formula is $L_\theta/L_z = (1 + 2\sin\theta)/3$ where θ is the angle of elevation above the horizon. The value of the zenithal luminance L_z depends on the angular elevation of the sun at the particular instant, but a conventional value is 2050 cd/ metre^2 which yields a conventional horizontal illuminance of 5000 lux; this may be a convenient value for some calculations but it is not a necessary assumption.

Daylight Factor (DF) The total daylight illuminance at a point (on a horizontal plane) inside a building comprises three components; direct from the sky, reflected from other buildings etc. outside, and reflected from room surfaces inside. Each of these three components is best expressed as a percentage ratio to the horizontal illuminance at an unobstructed point outside the building. More simply, daylight factor is the total amount of natural light reaching a reference point (on a horizontal working plane) within a building expressed as a percentage of the unobstructed external illuminance, both under a hemisphere of sky of known or assumed luminance distribution.

Sky Component (SC) That part of the total natural light reaching the reference point within the building directly from the sky without reflection from any surface, expressed as a percentage ratio of horizontal illuminances as before. It may be noted that *Sky factor* is sometimes calculated for a point in a building; this is the same as the sky component for a uniform sky but without any glazing in the windows or any external obstructions. It is a pure geometric rather than photometric ratio. Similarly, the ratio of window area to floor area is a rule-of-thumb which may bear little relation to the actual effective daylighting.

Externally Reflected Component (ERC) That part of the total natural light reaching the reference point within the building after reflection from surrounding buildings and the ground, expressed as a percentage ratio of horizontal illuminances as before.

Internally Reflected Component (IRC) That part of the total natural light reaching the reference point within the building after reflection from the various surfaces within the room containing the reference point, expressed as a percentage ratio of horizontal illuminances as before.

It follows that $DF = SC + ERC + IRC$. The value of SC depends on the geometry of the window, the transmittance of the glass and the luminance of the sky as seen through the window from the reference point; ERC depends also on the illuminances and reflectances of the external surfaces around the building; and IRC depends also on the illuminances and reflectances of the surrounding internal surfaces. It should be noted that the external and internal reflected light (ERC and IRC) are significant proportions of the total daylight (DF) in most urban buildings.

Measurements and predictions

We shall now consider how to measure or to calculate these components of daylight so that windows can be designed to give a desired performance in a proposed building or to check the performance of existing buildings.

An obvious approach would be to build a scale model of the room and its window and to make measurements outdoors under the natural sky. We immediately come up against the problem anticipated by Trotter, the variability of the natural sky, and the next obvious thing to do is to build an artificial sky. There are several ways of doing this.

At the Building Research Station and at several schools of architecture, this is a box with its vertical faces covered inside with flat mirrors, with its top made from a diffusing panel with lamps mounted above it, and with its base painted to simulate the ground. An aperture is provided in one side to accommodate the window wall of the model to be tested, unless of course the sky box can be made large enough for the whole model to be placed inside. At the BRS and at University College, London the model is placed on a table inside the sky box and adjacent buildings can be represented around it, but most sky boxes are about 4 ft square and 18 in. high. A close approximation to the CIE luminance distribution is possible if the reflectance of the base is carefully chosen in conjunction with a suitable diffusing material for the top and if the lamps are carefully placed.

There are several hemispherical skies, at Cambridge and at St

Helens in this country and in Moscow, Stockholm, Texas and Australia, which are much larger than the box skies. At St Helens, the sky is 22 ft diameter and sixteen 400 W MBFR lamps illuminate the white painted plaster surface to give the CIE distribution or fifty-six 80 W MCF tubes can be used to give uniform sky luminance. The Cambridge sky is made of translucent panels and the lamps are positioned outside it. In all these skies, the model has to be mounted centrally.

There are more than fifty methods of predictive calculation listed in the recent CIE Report 16 and three of these may be taken to illustrate the tabular method, the protractor method and the perspective method.

Tabular method The Simplified Daylight Tables give the sky component and the externally reflected component, for the CIE sky or a uniform sky. If the reference point is at distance d from the plane of the window, the height of the window head and the widths of the window sides to the left and right of the normal from the reference point can be expressed by the ratios H/d, W_1/d and W_2/d, for which the values of the components are tabulated. The sky component is the (algebraic) sum of its values for the two sides of the window; the externally reflected component may be estimated by splitting the window horizontally into two parts, one obstructed and the other not, and by assuming a modifying factor (typically 0·15) for the externally reflected component. Other tables are available relating room surface reflectances and proportions to give the internally reflected component, which has to be added to the first two to give the daylight factor.

Protractor method The BRS Protractors can be applied to scale drawings in plan and vertical section showing the windows in the side walls (glazed or unglazed) and showing roof lights which may be sloping or flat or vertical. It is only necessary to draw the sight lines from the chosen reference points, taking account of external obstructions, and then to apply the appropriate pair of protractors to the plan and vertical section drawings to get the sky and externally reflected components of daylight factor.

Perspective methods The basis is to draw the outline of the window and of external obstructions, as seen from the reference point inside the building, on an angular grid representing the quadrant of sky which contributes to the light through the window. In the Waldram diagram, the grid is spaced so that equal areas represent equal sky components all over the grid; the window sides plot as vertical lines but horizontal edges become slightly curved or sloping. In the Pilkington diagram, a pattern of dots is superimposed on the grid, each dot representing $0 \cdot 1$ per cent sky component, and a rectangular window or obstruction plots as a rectangle in true vertical perspective. In each method, the sky component and the externally reflected component are determined from the parts of the diagrams enclosed by the outlines of the window and of the obstructions, adding the areas on the Waldram diagram and counting the dots on the Pilkington diagram.

Internally reflected component The BRS Split-flux method is used with a horizontal plane dividing the room at mid-height of the window; the fluxes incident on the room surfaces above and below are assumed to be integrated by reflection and to be uniformly distributed so that the BRS tables or nomograms make the calculation easy. Alternatively, the Pilkington tables use the standard horizontal working plane and give the inter-reflected component appropriate to the position in the room. In both methods, it has to be remembered that the windows themselves may be regarded as having 15 per cent reflectance.

Interpretation of daylight factors

What use do we make of the calculated or measured total daylight factors at the various reference points in a building? Over the years since Trotter first measured his sill ratios and Waldram first calculated his sky factors, authoritative recommendations and standards for good daylighting practice have been evolved, using daylight factor as a measure of the lighting, as in the IES Code, the British Standard Codes of Practice and in various Government publications. These all give guidance on the minimum levels of daylight factor which, excluding direct sunlight, will ensure adequate natural lighting during daytime throughout most of the year.

But daylight factors are only a measure of quantity, not of quality. They tell us little or nothing about daylight glare from badly designed windows, about the appearance and modelling of objects within the building, nor even about the amount of lighting on sloping or vertical surfaces. These aspects of daylight design are no less important than quantity but they are less amenable to calculation or to model experiments. However, real progress is being made. The Daylight Glare Index developed by Hopkinson and Longmore has been described in admirable detail in the 1972 revision of IES Technical Report No 4 which amplifies much of the thinking in this chapter. The concepts of scalar illuminance and of illumination vectors arose largely from daylight studies and have been used powerfully in the analysis of the flow of light from windows, which is so different from and often so much preferred to the flow of light from ceiling-mounted artificial lighting. The designed-appearance techniques of J. M. Waldram can also be used to predict the way in which daylight reveals objects at various points within a building. It would seem that the designer, whether architect or engineer or whatever, has plenty of techniques available to ensure good design for natural lighting, if only he will use them. Some techniques are more complicated than others but even a simple method used intelligently can become a reliable guide to experience, particularly when combined with experience of simple methods of electric lighting prediction.

Le Corbusier said that in the history of architecture throughout the centuries, the history of windows has been an unceasing struggle between the desire for light and the laws of gravity.

Certainly, the design of buildings has been largely conditioned by the need for natural light and, if the economics of artificial light make this less important nowadays, we should use the established techniques to ensure high quality of daylighting and to enable the satisfactory integration of natural and artificial lighting to yield the best results from both.

12 What we want from Daylight

Chapter 11 dealt with the amount of daylight rather than with its quality; this chapter attempts to redress the balance.

Because electric light is so cheap, industrial and commercial managements may not be much concerned about the quantity of daylight; but nearly all of us demand those unmeasurable benefits of natural light that we call "quality", either the fleeting view of the weather outside, or the side lighting across a room, or the strong contrast of colour and brightness between the window and the walls. We want quality more than quantity, but we can't say how much more. Some of the disadvantages of daylight can be measured, such as glare or heat loss or solar gain, but we deliberately put up with these, or we spend money on controlling them, in order to enjoy its advantages.

The problem of quality

Quantitative design is reasonably straightforward; choose a convenient prediction method, apply it to the building and get the daylight factors at the points required. These values are compared with recommendations in codes or specifications and the quantitative acceptability of the design is settled, unambiguously.

When we come to design for good quality, we really run into difficulties. Quality has its impact on the emotions, so that aesthetic reactions are very much at a personal level and are therefore different from one person to another. Whether or not a particular aspect of a design will be liked by an individual using the building, or by another who is just looking at it, will depend very largely on his past experience, culture and training—it is simply not possible to please everybody. This does not mean that the designer can abdicate his responsibility for quality but it does offer some consolation when both praise and intense criticism are levelled at some completed project. We must also recognize that what might be generally regarded as good daylight design in one type of building could be the exact opposite in another which has a different application; the quality we want depends on who we are, where we are and on why we want it. What a specification! As an example, factories are often satisfactorily daylit by glass in the roof but an office which is daylit from the top may be a miserable place to work in. Architects know much better than lighting engineers that "good" daylight is not just "a lot of light".

Windowless buildings

It might almost be said that we like daylight for its own sake but we sometimes think we can get on equally well without it. Although fully artificial environments are undoubtedly the best in some instances, the arguments about them are highly emotional; it is shocking how little documented evidence there is to support the case for natural lighting. We hear terms such as view-out, contact, amenity, visual release, minimum glazing, preferred window shape —what do they all mean? How are they related or evaluated? If we knew, we could have a meaningful discussion of the merits of windowless buildings or of "minimum window buildings" (another unquantified requirement) instead of the usual economic analysis in which aesthetics and the like are utterly disregarded. In the cost-benefit assessment of windowless buildings, the costing side is clear but the benefit side is almost unmeasurable because we just don't know enough about it.

Analysis of quality

As implied above, properties which can be measured become known as quantities; those which cannot are considered to be qualities. For this reason, much of the window's contribution to the built environment is classed as qualitative; there is a marked difference in the "feel" of an interior lit from side windows as compared with normal electric lighting. The sources are in different planes and they provide light of different colours; one varies and the other is fixed, one transmits and the other emits. The visual effects are quite different and they determine the character of the typical daylit or electrically lit interior.

The main characteristic of daylight is that it comes through a window in the vertical plane at one side of a room. The light augments the illuminance on vertical surfaces, improves the definition of textures and gives a more pleasant modelling of objects. This flow of light across the objects in the interior might possibly be achieved with electric lighting but attempts have shown it to be difficult and expensive; it is virtually impossible with current off-the-shelf ranges of luminaires.

Secondly, the daylight from the window varies considerably in colour, which is usually regarded as a pleasantly acceptable feature. For critical colour-control tasks, north-facing windows may be used to eliminate direct sunshine during most of the working year but even the light from a north sky has enough variety to be interesting. The unpredictability of the amount of daylight from moment to moment or from month to month is at once its great advantage and disadvantage. The short term variations add an extra dimension to the daylit interior and undoubtedly contribute to the avoidance of monotony sometimes found under static installations. It is interesting that attempts to produce this effect in electric lighting seem strangely artificial and almost annoying.

The provision of contact with the outside world is perhaps the most important function of a window and is most often its prime justification. Man needs to be aware of the changes going on around him. He checks the time of day by clock; he can look out of the window for confirmation. He looks out also for release, in the form of movement compared with his static situation inside. He looks out of the window to check the weather and to reassure himself that life is still going on in the "real" world outside. In extremes, the window may prevent suffering from claustrophobia. No lighting engineer can even attempt to copy this important function of a window. In spite of all this, there is little hard knowledge about the purpose that is served by the view outside. There is some indication that the information content can be quite small; even a brick wall six feet away outside a window is much preferable to a brick wall at the same distance inside the same room. An open door or a glass partition between two offices may serve a similar purpose but somehow it isn't quite as good.

This is obviously a field for more research, following the lead of Hopkinson and Ne'eman who showed that the minimum acceptable size of a window depends largely on the information content of the view outside, rather than on the amount of daylight admitted or on the amount of electric light inside. In a narrow deep room, the preference is for a window extending the whole width of the end wall, but for wider rooms a window needs only to extend 50° or 60° in width; perhaps this is related to our limits of sharp binocular vision.

Finally, there is the quality aspect of quantity. Even in these days of prodigal electric lighting, there are still very many rooms which are primarily lit by daylight, in which we want enough light without excessive glare and without serious imbalance of heat. The compromise between advantages and disadvantages is still an assessment of quality even though the physical quantities are all known.

Daylight glare

Like any light source, the window is capable of producing glare, either disability glare or discomfort glare. As a broad generalization, disability glare is more usually caused by direct sunshine and discomfort glare by the blue sky or the clouds. Different techniques are required to overcome the two forms of glare.

We should note that "disability glare" means that the light impairs our vision of some of the things we want to see; "discomfort glare" is just uncomfortable and distracting but does not necessarily impair vision.

Disability glare from the sun may not always be unwelcome, because we like sunshine in these latitudes, but it does make things difficult to see and it is often associated with undesirable thermal problems. It requires shading devices which may be external sunbreaks or screens or internal blinds or curtains; visually these may be equally effective but thermally the external screens are to be preferred. Fixed shading devices can be accurately designed to exclude direct sunshine for selected periods of the year, usually horizontal screens for south-facing windows (in the northern hemisphere) or vertical screens for east- or west-facing windows. Careful design is in fact necessary because any errors in sun breaks or screens can be a prolific source of grumbles. Adjustable shading devices (blinds, curtains, slats and so on) can of course be brought into operation only when needed and therefore have less effect on the view-out function of the window. Again, careful installation can reduce maintenance troubles and can prevent avoidable grumbles.

It may be thought that the discomfort of glare from the sky would be best controlled by reducing the size of the offending window but this is not always the case; in some circumstances a small window can produce greater discomfort than a larger one. In such circumstances, a low-transmittance glass may be used to reduce the sky luminance without impairing the view-out function or the other subjective qualities of daylight. A large window with glass of say 60 per cent transmittance is likely to produce noticeably less discomfort glare than a window half the size but with clear glass, even though the amount of light is greater. The glass may be neutral grey or lightly tinted; either will be imperceptible from the inside because the eye will readily adapt to minor colour changes. The tinted glass may also affect the amount of internal electric light required to balance the daylight. There is scope for further research into this and into the merits of colour changes.

Discomfort glare can be quantified, by calculation rather than by measurement, in terms of the daylight glare index as described in the IES Technical Report No 4 "Daytime lighting in buildings". Although this is similar to the IES Glare Index for electric lighting installations, the two indices are not related and cannot be combined by any general formula. We have no accepted way of calculating discomfort glare from windows and from electric luminaires at the same time. The degree of glare is greatly influenced by the detailed design of the interior and particularly of the window wall; splayed reveals help to reduce the hard contrast and a light decorative treatment on the floor and the window wall will provide reflected light that reduces discomfort. In general, discomfort glare is more prevalent than we sometimes realize and it can be greatly alleviated by ensuring that the room is bright and clean; if high daylight factors are required from side windows, there is much to be said for having windows in two adjacent walls so that both windows walls are directly lit and the adaptation level of the eye is raised.

Natural and artificial lighting

It is curious that natural lighting has been closely studied for a hundred years, resulting in scores of text books and learned papers, and that electric lighting has been studied for over 50 years, with comparable documentation, but neither of these really serve our present needs because we like to have them both at once. We used to have natural light in daytime and artificial light at night; we still design our buildings and our lighting installations as though these sources of light were used quite separately. There have been notable exceptions to this disconnection of thought, under labels such as PSALI (Permanent supplementary lighting of interiors), "heat through light" and "integrated design", but these have not really penetrated into the run-of-the-mill lighting schemes. Electric lighting is usually designed for night-time use and is then switched on all of every day; windows are usually designed to admit daylight and are then used mostly for view-out during the day and to show the light at night. We need to understand the functions of these two closely related features of our buildings and to learn to link them together in our everyday thinking.

Rooms such as offices may be considered in three groups: those not more than 5 m or 6 m depth from the window in which daylight may be expected to be adequate for much of the working day; those up to 10 m or 12 m depth in which the character of daylight can be retained if it is properly supported by electric light; and those much deeper rooms in which daylighting is never adequate and permanent electric lighting is necessary.

In the first group, the electric lighting is unlikely to be used during daytime often enough to justify the installation of lights specially for daytime use. It is therefore sufficient to design the installation for night-time use, taking care that the colour of the light and the light distribution from the luminaires will give an acceptable blending when the lamps are switched on during a dull period of daylight.

The second group requires that the daylight character should be preserved and that an artificial supplement should be provided because the depth of the room is too great for adequate daylight penetration. The two forms of lighting have complementary roles. The daylight provides the flow of light, improving the modelling of objects and illuminating vertical surfaces. Part of the electric lighting installation should be so designed as to provide adequate illuminances towards the back of the room, without swamping the directional properties of the light from the window. The other part of the installation should be designed for night-time use over the whole area, without regard for daylight but taking account of the office layout which will probably be related to the daytime flow of light from the window. The illuminance level given by this night-time installation may be less than that of the supplementary lighting provided at the back of the room in daytime. The two parts will usually have some luminaires which are common to both, so that a carefully planned array of suitable luminaires will serve both requirements by means of selective switching. Human nature being what it is, the switching should be clock-controlled, with manual over-ride, to ensure that the balance of illumination is maintained.

It may be noted that the natural and artificial lighting systems involved in this second group ought properly to be designed together, in terms of window size and height as well as room size. Particularly, a window glazed with tinted glass may prove an economy in operating costs because lower levels of supplementary light would be adequate to balance the reduced natural light in daytime.

In the third group, the window serves primarily the view-out function and does not provide any significant illuminance or flow of light over much of the working area; the electric lighting installation becomes much more important and calls for greater attention to the detail of its design. The 1973 IES Code recommends a standard service illuminance of 500–750 lux for general office installations but this may be increased either by some factor considered in the Flow Chart (on page 48 of the Code) or to ensure a proper balance with the natural light in day time; many modern offices start life with an initial (new and clean) illuminance up to double the standard Code recommendation. With this large amount of light, and the sophisticated equipment and heavy costs associated with it, a wise designer should try to introduce some of the benefits of natural lighting already mentioned. This is a matter for the flair and skill of the designer, but it may include non-uniformity in the spacing or in the colour or in the type of luminaire, all in planned variety. We have in past years acknowledged the genius of some architects in designing windows to suit the building methods of their time; now that the engineering of buildings has released many of the constraints and the idea of the "planned environment" includes light as well as heat, sound and ventilation, is it too much to expect that some fresh genius will appear even in office lighting?

As we have said before, the lighting engineer has much to learn from his daylighting colleagues in the architectural profession. It has been the object of these two chapters to describe some of the basic thoughts involved in the use of daylight inside buildings and it is to be hoped that readers may study the benefits of daylight more closely and may examine how our monotonous installations might be enlivened. If we have pointed to some areas of ignorance, perhaps this will stimulate enquiry or research into the modern fashion of having windows and lamps at the same time.

13 Hours of Sunshine

"In the centre of everything is the sun. Nor could anyone have placed this luminary at any other, better point." So wrote Copernicus in the 16th century. However, present-day students of the sun and its movement find their task much simpler if they follow the theories of Aristotle and Ptolemy and regard the sun as travelling round a fixed earth. We will do the same, and will choose convenient diagrams to represent the apparent movement of the sun. These diagrams present few difficulties to those already familiar with three-dimensional representation, such as daylight factor protractors or iso-candela curves. We shall consider the northern hemisphere at the latitude of London ($51\frac{1}{2}°$N).

The sun's movements

How does the sun move across the sky? The arc of the sun's path is low in winter and extends only from SE to SW; in summer it is high and extends from NE through S to NW. Fig. 13.1 illustrates the movements of the sun at the mid-summer solstice, at the spring and autumn equinoxes and at the mid-winter solstice, showing the changing elevation and azimuth angles of the sun in relation to the seasons. This diagram was drawn for 45° latitude for convenience's sake. Each oblique ellipse in Fig. 13.1 represents 24 h, so this diagram also shows the durations of daylight (above the shaded horizontal plane) and of night—the dashed lines below the horizontal plane indicate, for example, the long twilight of summer or the dark nights of winter.

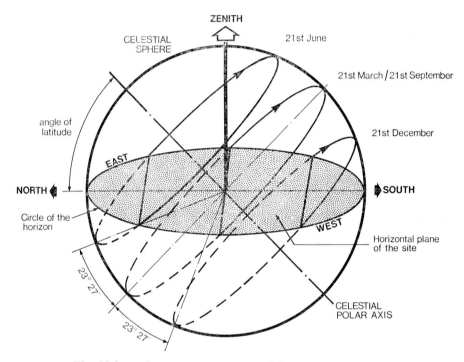

Fig. 13.1 Apparent movements of the sun

From Fig. 13.1, we can deduce some interesting facts and correct some popular misbeliefs:
(i) In our latitudes, the highest elevation of the sun (at the solar

noon of any day) is always due south. The noon elevation is a maximum on June 21 and a minimum on December 21, the actual values being $(90-51\frac{1}{2}) + 23\frac{1}{2} = 62°$ in summer and $(90-51\frac{1}{2}) - 23\frac{1}{2} = 15°$ in winter. Sunrise is SE in mid-winter, E at the equinoxes and NE in mid-summer.

(ii) In the near-equatorial regions of the world, between the tropics of Cancer and Capricorn ($23\frac{1}{2}°$N and $23\frac{1}{2}°$S), the noon sun may be due North or due South, depending on the season of the year. The sun passes through the actual zenith only twice a year; at the equator this occurs only on June 21 and December 21, and not every day as popularly thought.

(iii) In the near-polar regions, above the Arctic Circle ($66\frac{1}{2}°$N), the sun never sets during some periods of the year but makes a complete circuit of the horizon which inevitably makes it pass north. At the North Pole, the sun would be visible (in clear weather) continuously from March 21 to September 21 and invisible for the rest of the year.

(iv) In all parts of the world other than the poles, the sun rises exactly east and sets exactly west on only two days of the whole year, namely the equinoxes March 21 and September 21.

The sundial

Shadows move with the same precision as the sun, and the changing trace of a shadow on the ground or on a wall has been used for time measurements for over 3000 years. Essentially, the style or gnomon of a sundial has a straight edge parallel to the polar axis of the earth, so that the position of its shadow depends only on the rotation of the earth and not on the seasonal changes in the elevation of the sun's path. The plane of the dial may be in any orientation, with the graduations of the hour scale spaced accordingly. The shadow will always be due north of the style at solar noon, and, for other times, the graduations may be determined by experiment with a reliable watch or by calculation.

Owing to the effects of the earth's orbit round the sun, the apparent solar time differs somewhat from the mean siderial time, with a maximum discrepancy of 16 min in November and February but with negligible difference on April 15, June 15, September 1, and December 24. This does not affect the calculations of the duration of sunshine.

Sunpath diagrams

Direct sun penetration through the windows into a building can be regarded visually as an amenity or a nuisance. A sunpath diagram may be used to determine the directions and duration of penetration, so that the degree of amenity or nuisance can be assessed. The representation of the hemisphere of sky as a plane diagram can be achieved in many ways, well known to map makers, and the direction of the rays from the sun at any particular instant can be plotted in terms of its angles of azimuth and elevation. Figs. 13.2 and 13.3 are convenient forms of sunpath diagram, and serve to illustrate the method of using them.

Fig. 13.2 Sunpaths for south elevation in latitude $51\frac{1}{2}°$ north (equatorial gnomonic or perspective projection)

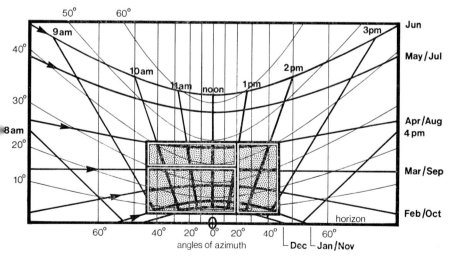

Fig. 13.2 is analogous to a south-facing elevation view. It has a background grid of vertical lines each representing an angle in azimuth from due south as zero (around the horizon) and of curved lines each representing an angle in elevation (above the horizon). This is an "equatorial gnomonic" projection, which has the advantage that rectangular objects are represented by rectangles on the diagram in ordinary perspective. The bold curved lines are the paths of the sun at each month of the year, from December to June and back to January, with the hours of the day shown as bold straight lines. Note that if any other azimuth than south were to be taken as the zero, that is for a building aspect in any other direction, the bold lines would be arranged differently.

A reference point is chosen within the building to be studied, and from this point the co-ordinates of the edges of the window are plotted on the same background grid; similarly, the co-ordinates of any external obstructions (buildings, shading devices, etc.), so that the diagram represents the view out of the window from the reference point. The sunpaths are then superimposed and Fig. 13.2 shows the result.

From this composite diagram, the portions of the sunpaths within the window outline show the times of day and of year when direct penetration reaches the chosen reference point as follows:

December	0940–1420 h
January and November	0850–1510 h
February and October	0910–1450 h
March and September	0930–1430 h
April and August	1020–1340 h
May to July	none

Thus a worker in a south-facing office with such a window might receive direct sunshine during the times shown above. The building designer has to decide whether this would be regarded as an amenity to be welcomed or a nuisance to be avoided (by blinds, screens, absorbing glass etc); this decision is based on the designer's

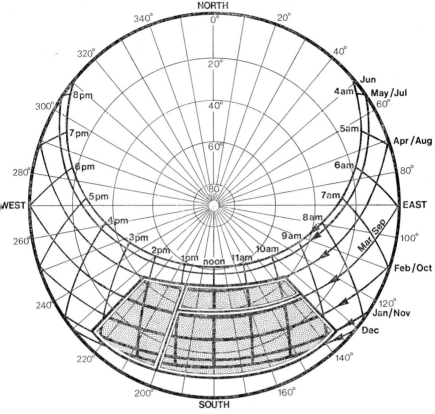

Fig. 13.3   Sunpaths in latitude $51\frac{1}{2}°$ north (zenithal equidistant projection)

knowledge of the types of work, periods of usage etc. for this particular situation.

Fig. 13.3 shows the result of the same analytical process on a diagram which is analogous to a plan view. The background radial lines represent angles in azimuth, and the circular lines represent angles in elevation, with the zenith in the middle. The bold lines are the sunpaths for $51\frac{1}{2}°N$ latitude and the hour lines are as before. This is a "zenithal equidistant" projection, which has the advantage that it can be used for a building with any aspect direction. The plotting of the window area is, however, somewhat less easy than in Fig. 13.2, but the results are the same in terms of the times of day and of year.

Probability of sunshine

The potential duration of sunshine at latitude $51\frac{1}{2}°$N varies between $7\frac{1}{2}$ h per day in mid-winter and $16\frac{1}{2}$ h per day in mid-summer. The actual duration is less because of cloud, averaging about 15 per cent in the winter months (1 h per day at mid-winter) or about 45 per cent in the summer months (7 h per day at mid-summer). The effective duration is less still, because of interference by other buildings or natural objects and because of the greater distance through the atmosphere that the sun's rays must travel at low angles. If the effective horizon is taken at $10°$ elevation, the average effective durations are reduced to about one-third in the winter months (20 min per day at mid-winter) or to about five-sixths in the summer (6 h per day at mid-summer.)

Building design

The exercise illustrated in Figs. 13.2 and 13.3, with the addition of the probability calculations, may be necessary to determine how well a building design performs in terms of certain planning guides, or to indicate what protection may be desirable for particular windows or work places. The penetration of sunlight into interiors gives contact with the outside and provides an ever-changing pattern of light; in this respect it has great advantages over artificial light. The possible disadvantages of sunlight, objectionable glare or excessive heat gain, are capable of being worse than in artificial light. The careful study of "insolation" is thus a part of the whole study of illumination.

Further reading

"The science of daylight", J. W. T. Walsh (Macdonald, London, 1961)

"The sun", P. Burberry, *Architects' Journal*, **143** (Jan. 1966)

"Sunlight and daylight" (Department of the Environment, 1971)

"Sunlight" in BS CP3 "Basic data for the design of buildings", Chap. 1 Pt. 3 (British Standards Institution, to be published in 1975)

14 Road Lighting I

We used to call it "Street lighting" but the revised British Standard Code of Practice 1004, published in September 1973, calls it "Road lighting". Quite rightly so, because the lighting of country and suburban roads has become no less important than that of a city street.

Road lighting is perhaps the most closely controlled installation in our daily experience; controlled in finance and authorization, controlled in design of installation and of lantern, controlled in amount and performance, controlled in beam shape and beam direction. Although we used to complain about the differences between one length of road and the next, we now find a remarkable uniformity in the quality of recent installations all over the country. It is also surprisingly efficient and effective. Our thanks to the Department of the Environment (which incorporated the old Ministry of Transport), to the BSI and to the APLE.

It breaks all the usual rules of lighting. The general rule mentioned in chapter 2, namely, "Lighting from the right place, to the right place, at the right brightness and colour", is still obeyed but the "right" places are unusual by interior lighting practice. The lanterns are put a long way ahead of us, shining their beams towards us and causing glare that would usually be unacceptable. The light shines on to the road surface and not on the obstructions that we want to see, leaving them dark and emphasizing their shadows—we call it silhouette vision. The amount of light is pitifully small, between 0·2 lux on a minor residential road and 20 lux on an average main road. The colour of the light from the most popular lamps, low pressure sodium lamps, is an abominably monochromatic yellow such that pillar boxes, grass and panda cars all look brown or colourless.

But it works. It works so well that a 90 W lamp will give good lighting for a 30 m (100 ft) length of a busy main road, and motorists will switch off their headlights and still have 30–40 per cent less accidents than if the lighting were bad. It works so well that the basic principles described in the 1937 Departmental Report are still valid in 1975 in spite of new kinds of lamp, completely new kinds of road surface and a new generation of public lighting engineers. It is a British pride that although our best may not be as good as the best in some other countries, the general run of road lighting in this country has no equal in the world.

How it works

Look at a single street lamp from a distance. There is a bright patch of light reflected from the road surface, extending from below the lamp (and not from beyond it) to about three-quarters of the way towards you. It also extends partly across the road near the lamp, becoming narrower towards you so that when seen in perspective it is a T-shaped patch.

Imagine a pedestrian walking along the road out of the distant darkness. He will be invisible until he gets near the lamp when for a few seconds he will be seen clearly in the direct light from the lamp, while he is still approaching it. Then suddenly he is seen in clear silhouette, as a dark figure with first his feet and then his whole body seen against the bright reflected patch on the road, and this continues until he is close to you.

Now consider all the lamps along the road. They are so placed that the bright T-patches merge together and coalesce into a uniformly bright (or nearly uniform) surface. The pedestrian can be seen in silhouette perhaps a quarter of a mile away (assuming, with every road theoretician, that the road is straight and level) and he will be seen consistently in silhouette as he walks all the way towards you. This consistency in appearance is one of the secrets of effective lighting; another is the relatively high brightness or luminance of the road surface and the hard contrast of the obstructions on the road. As the pedestrian approaches each lamp along the road and its light falls on his face, a transient change occurs in his appearance—however, it seems quite natural and understandable that you should see his face when he is close to the lamp and you scarcely notice this fleeting reversal of silhouette vision; the consistency in his appearance seems to be maintained. This, in simple terms, is how we see obstructions on our roads, not by the light falling on them but by the light falling on the road surface beyond and around them.

In more complex terms, we see also the shadows cast by the obstructions whether they are moving or fixed, we see the glint of reflections from shiny surfaces, the texture of light and shade contained within the obstruction (we call it "modelling" in interior lighting), the colour differences and the effects of illuminated haze; we are particularly sensitive to movement which causes changes in these features. So we recognize the obstruction, realize what is happening and take the action necessary to avoid trouble. We have learned, by experience rather than by teaching, to do all this quickly and accurately in light which is too dim to read a newspaper—a remarkable combination of efficient lighting and a concentrated discipline.

White, blue or yellow light

Mercury lamps have a fluorescent coating nowadays and are white rather than the bluish-green of 20 years ago. Gas lamps and some metal halide lamps give a somewhat greenish white light but these are a small proportion of the total. Fluorescent tubular lamps give a variety of white lights whilst tungsten filament and high pressure sodium lamps give a yellowish white; all these nominally white lamps seem to be equally effective in respect of their colour and permit us to recognize obstructions, etc., in terms of their known colours. This is an aid to detection and visibility, by presenting the obstruction in familiar colour. The author must acknowledge a personal bias against the monochromatic yellow of low pressure sodium lamps even though they are otherwise admirably suited to the lighting of all kinds of road. The loss of colour sensations is a deprivation which is generally regarded as quite acceptable for heavily trafficked roads and which is compensated by the undoubted economies provided by these lamps, permitting greater mileages of road to be lit for the same money. Perhaps the colours of other lights in the road scene are appreciated better, such as signs, signals, vehicle lights, shops and so on which provide us with information by which we can see our way more safely.

Light distribution

The differences between lamps are not only their colour. Their size and shape affect the design of lantern to make best use of their light; their efficacy and life affect the whole economics of the installation. Filament lamps and tubular fluorescent lamps are disappearing except from residential roads where their colour rendering properties have advantages. For traffic routes, the combination of lamp and lantern must so control the light as to extend it as far as possible along the road without causing glare and as much across the road as need be; this leads to horizontal-burning lamps to give the optimum control in the vertical sense.

Although the geometrical layout of lanterns and columns is fairly well established and fully described in the several parts of the Code of Practice, there is still much scope for skilled development in the design of light distributions. The principal objective is to achieve a high and uniform luminance of the reflected light from the road surface, but road surfaces vary enormously from day to day as well as from one street to the next. It is therefore impossible to determine the ideal distribution of light intensity from the lanterns. The present trend in the changing properties of road surfaces seems to call for lanterns with lower and wider beams, which in turn calls for closer spacing or greater height of the columns and places more importance on the reflective properties of the walls, pavements, verges, etc., along the sides of the road.

The present geometry of installations is based on some basic formulae developed in the late 1950's, before much was known about the reflecting properties of modern anti-skid road surfaces. The formulae were empirical, reinforced by calculation of perspective viewing; that is to say they recorded the best practical experience and were checked to ensure that their results were consistently reasonable, which is a very good way of making rules. It is gratifying that they give results closely compatible with the most advanced modern techniques of calculation and measurement. This is no place for the detailed formulae, which are legion, but we may note that they relate the spacing/height ratio to the type of lantern, the arrangement of the columns, the width of the road and the group or quantity measure of the lighting installation. All this from a simple table of figures in the BS CP 1004, without recourse to any computer, and the result is as close as need be to the most elaborate calculations.

One of the secrets of the success of this design technique is the consistency in performance of the lanterns, governed by British

Standard 1788 which defines the relative "shape" of the light intensity distribution rather than the absolute values. This was quite new ten years ago but it has not been changed much and will probably continue in the next issue, which is a compliment to those who produced it and to those who insisted that it should be observed. Further, the successful control of glare within limits which are acceptable to British road-users has enabled the great increases in lamp efficacy (up to 50 per cent in the last ten years) to be applied to the improvement of the standard of road lighting without any changes in the basic techniques.

Reduction in glare tends to bring reduction in uniformity of the luminance of the road surface; in extreme cases of widely spaced cut-off lanterns, a "ladder" effect of transverse bands may be produced along the road. This has long been a matter of debate, some taking the view that the effect is only pronounced at some distance ahead where any obstruction would be seen against several "steps" of the ladder and would therefore be seen clearly, and others taking the view that the high contrast of silhouette vision is impaired if the background is not uniform. More insidious is the blotchy effect that may be produced by bad camber or bad road surface or bad light distribution; this may be unnoticed by a vehicle driver and may so reduce the contrast of an obstruction that the danger is also unnoticed. One of the features of the standard semi-cut-off light distribution is that it balances the disadvantage of glare against the advantage of longitudinal uniformity, with the proper emphasis on the latter.

Wet roads

The theory and practice of road lighting are nearly always based on dry roads, even though the added dangers of a wet road call for better lighting. The Danish authorities are attempting to design for wet roads and their success will be watched closely. The T-shaped patch described earlier becomes narrower when the road is wet, so that the patches from the several lanterns may not overlap and dark longitudinal streaks may appear along the road. Fortunately, the rather shiny dry road surfaces of the 1950's were not unlike modern anti-skid road surfaces when wet and the same lighting techniques can be applied. It is, however, very important that road surfacing materials should be chosen, by the civil engineer or his maintenance foreman, to suit the conditions provided for lighting them in the wet as well as in the dry; this means in general that rough, quick draining materials are desirable, both for skid resistance and uniformity of lighting.

New techniques

As light has become cheaper, other methods of road lighting are being examined. The established process of lighting the road surface as a bright background against which obstructions are seen in high contrast involves putting the lanterns in exactly the right places; this is simple along a straight and level road. At bends, the lanterns must be closer and on the outer side; at humps or at the top of a hill, they may cause glare and also may fail to brighten the road surface adequately; at junctions or roundabouts, the additional columns add to the collision hazards; at multi-level junctions, the arrangements of the lanterns appear quite chaotic and may fail to give proper optical guidance; in short, there is a need for particular solutions to particular problems.

High masts, twice or three times the height of ordinary columns, are very suitable for complex and multi-level junctions, providing light over the whole area from a greatly reduced number of columns. Their positions must still be chosen carefully, to provide adequate road surface luminance, to avoid shadows and to give some crosslighting so that reasonably uniform conditions are maintained; the lighting engineer will have to work with his civil engineering colleague to make sure of having foundations in the right places. The lighting from high masts, like that at roundabouts from ordinary levels, is more akin to floodlighting than to silhouette vision and more light is needed for an equivalent visual effect, but this is fully justified because the visual tasks in such locations are more complex than on straight roads. High mast lighting may not be justified if the area to be lit is long and narrow or includes a large proportion of unused ground such that it would be better lit by traditional methods.

Another technique which we have largely imported from the Continent is "catenary lighting", or more accurately, "axial median lighting". This comprises a row of lanterns, usually suspended at close spacing/height ratio along the central reserve of a double carriageway, directing their light more across the road than along it. The lanterns do not hang over the road but may be perhaps three metres behind the kerb, and they must give a wide spread of light so that there is a considerable oblique flow of light to produce a reasonable road surface luminance. Such an installation gives very good longitudinal uniformity, which is one of its advantages, whilst the reduced luminance near the outer kerb is compensated by increased illumination of the verges, etc. Vision is still largely by silhouette, because the light does not fall strongly on the front or back of any obstruction, but there is enough of the floodlighting effect for obstructions to be recognized easily. Again the amount of light should be more than in the traditional technique.

It may be noted that the cross-flow of light, across or oblique to the road, is a beneficial feature of both high mast and axial median lighting and this is becoming recognized as a desirable feature in traditional road lighting. This oblique flow was given by the early lanterns with low pressure sodium lamps, because of their length and the consequent wide spread of the beam from the lanterns; perhaps this imponderable effect was one of the reasons for the success of these lanterns.

Some special requirements of roads which are not just straight and level and without junctions are discussed in the next chapter.

15 Road Lighting II

Although the primary justification for road lighting is the prevention or reduction of traffic accidents and of crime, most of us regard road lighting also as a contribution to our convenience and to our speed of travel; it gives us the confidence that we want. The engineering of road lighting is, however, still concerned with the cold figures of statistics and performance rather than with amenity or comfort. This follows a deliberate decision to light as many miles as is economically possible and to reduce accidents as a first priority. The result is a very careful cost/benefit assessment of a just-tolerable visual standard for a just-tolerable cost.

How much light?

Having incurred the capital costs of an installation, we can obtain greater benefit by operating it all through the night; it follows that high efficacy in the lamps and high utilization in the lanterns are very desirable to make best use of the electric power consumed. Lamp efficacy is easily measured, which is one reason for choosing low pressure sodium lamps wherever their colour rendering deficiencies can be accepted. Lantern utilization is more difficult but is simply assessed in current British Codes in terms of the downward light output ratio for Group A lanterns or total light output ratio for Group B. The true utilization factor (lumens on carriageway/lumens from lamp) is not often considered because the light on the sides of the carriageway is no less important than the light on the carriageway itself. We design for the carriageway only but we take care that there is some extra light spread along its sides, on the footpath or the hedges or the buildings or whatever. This is fundamental in current British practice.

Lamps and lanterns have greatly improved and the spacing/ height ratios have been reduced in the past 20 years, with the result that the amount of light on the road has increased by a ratio of about 2·5. In the same period, however, the rougher non-skid surfaces have more than swallowed up the extra light, so that our present road surface luminances are no greater than under the early installations to the 1952 Code. We can, however, see much better, because the luminance of the road surface is more uniform, the sides of the road are brighter and the objects on the road (people, vehicles, obstructions, etc.) are better illuminated. Although our engineering designs are based on the light actually reflected from the surface of the road, the great improvements during the past 20 years are mostly in the light which does not reach the road surface.

The 1952 issue of BS CP 1004 was fully revised in 1963 for Traffic Routes, using the same principles and formulae as were described in Chapter 14; the 1974 revision of Pt 2 has much the same formulae adapted to modern road surfaces. Very roughly, typical figures for main road illuminances are 15 lux for semi-cut-off lanterns or 20 lux for cut-off lanterns, giving a luminance of about 1·5 candelas per square metre or 0·5 foot lamberts. The luminance figures vary widely, partly because the illuminance/ luminance ratio ranges between five and 25 even for main roads according to the degree of polish and to the light distribution from the lanterns, although the actual amount of light is fairly consistent. Group B lighting installations (not to be confused with B roads) are for lightly trafficked roads and have much less light. BS CP 1004 Pt 3 (1969) described several alternatives, from 6 m columns at 40 m spacing to 13 ft columns (nearly 4 m) at 60 m spacing; the latter are more common and the average illuminances range from about 3 lux to 0·2 lux or even less. These are really guidance lights, to show the line of the road rather than to illuminate obstructions on it; motorists who drive on such roads without headlights are foolhardy.

To summarize, neither average illuminance nor road surface luminance is a proper measure of road lighting because the difference between good and poor lighting is more than quantity alone.

Bridges and humps

A bridge or an elevated road has two particularly difficult features —there is nothing much beyond the kerbs to provide visible surroundings, and the results of a driver's failure may be catastrophic. A long bridge is usually humped and the distant lights may be seen as a cluster just above the crest of the road. where they may conceal an obstruction rather than reveal it. Add to this the architect's proper resistance to unwanted columns and the civil engineer's difficulty in providing foundations for the columns; it may be said that bridge lighting is an art rather than an established practice. Each installation must be tailored to suit the bridge and the advice in BS CP 1004 Pt 6 is guidance rather than a prescription.

The first essential is to light the parapets, delineating the boundaries and extending the background as widely as possible. Secondly deeply cut-off lanterns should be used to avoid glare and to prevent the occurrence of any confusing pattern of lights. Thirdly, sufficient cross-lighting is required to illuminate any obstructions and to model them strongly. Fourthly, the spacing must be close enough, giving high uniformity to avoid pools of shadow. The list could be extended but this is enough to show that long bridges have their own problems. Fortunately, these problems are simplified by the freedom from cross-roads and the tradition of careful discipline for the driver.

Short bridges are like any other length of road, but, if high columns are precluded for aesthetic reasons, more light will be needed at each end.

Junctions and roundabouts

The lighting design for urban streets usually starts at the junctions, putting the lanterns in the right places in each approach road and then filling the intervening stretches with the appropriate arrangement. Closer spacing at the junctions will provide more light and better modelling and will help to maintain the uniformity. The surroundings of the junction, whether buildings or pavements or pedestrian barriers, will receive more light than on a simple straight road.

This arises from the practical experience that hazardous conditions are more likely to originate somewhere off the road than from the carriageway ahead, and there are many more potential hazards around a road junction than on a straight road. The techniques described in detail in BS CP 1004 Pts 4 and 5 are based primarily on the need for seeing other vehicles silhouetted against the pattern of roads but fortunately they also provide extra light on the areas surrounding the junction.

The first objective is to reveal the existence and formation of the junction and to ensure that the arrangement of lights does not give a misleading impression of the run of the roads. At a T-junction, a lamp straight ahead on the line of the approach alerts the driver but, if this were done at a roundabout, this particular warning might be lost amongst all the other lights; experience has shown that a "target" lantern on a large central island (as in

BS CP 1004 of 1952) may itself be a danger. Secondly, a near-side light just around the corner of a left-hand turn serves to silhouette a vehicle which is entering the junction and to illuminate any obstruction that was invisible to its driver before making the turn. Similar reasoning applies to a nearside light on the opposite side of a cross-road. In the normal development of the installation, this near-side light would be followed by an off-side light somewhat farther away from the junction. At a roundabout there would be a succession of near-side lights around the outer kerb placed so as to reveal other vehicles or obstructions farther away, following the usual rule of putting lights on the outside of a bend.

Again we see the need for putting lanterns in the right places to provide a luminous background for silhouette vision as far as this is appropriate and to provide the cross-flow of light to help perception and recognition of obstructions. The flow of light on to the objects themselves is no less important than the reflection of light from the road surface.

Roundabouts come in many sizes and with great variations in traffic flow. Large roundabouts have a ring of lanterns, arranged round the outer kerb and perhaps on the splitter islands in the approach roads, giving 20 lux or more at either kerb and also serving to announce the presence of the roundabout from a distance. This specially arranged lighting is necessary because a roundabout is an obstruction in the flow of traffic, with turning and weaving manoeuvres, and the lighting must be continued for 150 m or so along each exit road to help the traffic to sort itself out safety. No lanterns should be mounted on high columns on the central island because they would confuse the ring shape and also because the columns would be a collision hazard.

Small roundabouts, with islands less than about 15 m diameter, may be lit by a symmetrical lantern on a central column if there is also adequate lighting along the approach and exit roads but this must depend on traffic density.

Very small roundabouts are a relatively recent feature of heavily trafficked roads; they are used to ensure safe "hooking" turns (off-side to off-side for each vehicle) at a junction which might otherwise require traffic signals. Some have raised mushroom islands, over 4 m diameter, which have to be circumnavigated and some have painted "spots", usually less than 4 m diameter, which may be overrun by long vehicles. The closely knit traffic manocuvres require very carefully designed lighting and we may hope for glare-free floodlighting up to about 50 lux, with corresponding lighting for the approach roads. Channelized junctions are of the same general type but with double white lines and hatched areas to control the positions of vehicles; here we may have stationary vehicles being passed by others at high speed and it is particularly necessary to ensure that the amount and character of the lighting are sufficient to reduce the hazard.

Pedestrian crossings and white lines

The flashing yellow beacons at each end of a pedestrian crossing do not illuminate the pedestrians but serve only as a warning to vehicle drivers. Their value is less than it was originally because there are now so many other flashing yellow lights about. The pedestrians must therefore be revealed by proper arrangement of the road lighting installation and this should take precedence over its usual function of lighting the road. This requires that the pedestrian crossing should be at mid-span, to give symmetrical lighting, with the nearer light on the left of a pedestrian when he starts to cross, so that he is revealed in silhouette to traffic approaching from his right. This may be insufficient in heavy traffic and supplementary floodlighting may be installed to emphasize the crossing, to illuminate pedestrians waiting on the pavement and to provide a further 50 lux on the crossing.

The white stripes of pedestrian crossings may look very bold in daylight but the contrast is reduced at night when the motorist looks at them at a near-glancing angle. The pedestrian may be unaware of this difference. The white material should be somewhat shiny although not slippery; the material of the black stripes should be as dark as practicable and, more important, should have a harsh matt surface to reduce specular reflection, particularly when wet. In practical experience this is achieved nowadays by some Local Authorities; perhaps this is an outstanding but rare example of success by lighting and safety engineers in insisting that their civil engineering colleagues should provide a road surface material which has the right reflection characteristics when wet or dry.

All white markings on the road surface tend to lose contrast at night and, if they are not of the right material, there is little that the lighting installation can do to reveal them. Reflectorized white paint has considerable merit when it is in good condition and when the principal lighting is the motorist's own headlamps, but its value tends to be reduced by a generous road lighting installation. A raised kerb can be revealed by ensuring that the riser is in shadow; a sloping kerb may, however, cause the reverse effect and show as a bright line which may give quite the wrong impression.

Underpasses and tunnels

Underpasses or bridged roads may not require lighting in daytime if the covered length is less than 50 m, or after dark if less than 15 m; both figures are approximate. Greater lengths are insufficiently lit by the diffusion of light from their ends and they call for carefully designed installations on a more generous scale than open roads. Fortunately, underpasses are often so expensive to construct that the cost of proper lighting is proportionately low, both in comparison with the main structure and with the possible cost of accidents in it.

Any usual night-time installation is of little value in daytime, although it may give enough light for pedestrians because they have time to adapt. For fast-moving vehicles, the amount of light must be very much greater, commensurate with daylight.

The visual problem in a tunnel is first for an approaching driver to be enabled to see sufficiently far into the "black hole" while he is still outside in the daylight; secondly for a driver within the tunnel to be enabled to see far enough ahead as he gradually adapts to the lower luminance levels. This means a delicate balance between speed and contrast on one hand and cost on the other— also a shrewd estimate of the likelihood of a driver slowing down if he cannot see properly. For example, safe vision in moderate daylight requires that the average luminance within the portal of a tunnel should be 200 cd/metre^2, requiring about 2000 lux for a distance of about 60 m to suit a speed of 80 km/h (50 mph). The extra cost of providing wall surfacing materials of high reflectance and of keeping them clean has to be balanced against the high cost of illumination—a long tunnel is usually wisely designed in this way but short tunnels may be so neglected as to become Stygian.

This discussion emphasizes the enormous difference between the brightness levels in daylight and under road lighting installations at night, a much greater difference than we experience in interior lighting in offices and the like. When road users start to demand more light or higher reflectances, the character of our roads may change; until then we can be grateful that the lighting industries of the world have achieved such success with so little light.

16 Foggy Days and Misty Nights

Seeing through fog is difficult at any time, and is so complicated at night that we cannot examine the details in a short chapter. The major effect of fog is to reduce the contrast of colour and brightness among the things around us. There are all sorts of clever devices from lasers to tinted spectacles that are supposed to help us to see, but unfortunately they are all subject to the classical laws of optics and are equally blanketed by the fog, whatever their inventors' claims.

In lighting engineering, we expect the air to be clear and we rarely take account of atmospheric scatter, although it is mentioned in IES Technical Report No 9 on "Maintenance". However, anyone concerned with all-weather outdoor lighting, and particularly with long-range lighting or signalling, soon realizes that he must consider the properties of the atmosphere through which the light is transmitted.

Clean fogs, that is without dust or smoke particles, do not absorb any light, because they consist of pure water droplets suspended in clear air. They scatter light, they reflect it and diffuse it and generally prevent it from going where we require it. This scattering has three effects: to attenuate a beam and thus to reduce its intensity, to diffuse the light and thus to reduce the shadows, and to fill the air with a luminous veil and this to reduce contrast. These effects are really the same, because the attenuation is due to scattering of light which is then diffused to form the veil of "air light", but the third effect, the reduction of contrast, is usually much the most important.

Fog can occur when humid air is cooled so that it becomes supersaturated. If it also contains the right sort of nuclei, tiny droplets are formed and stay in suspension in the air until they evaporate or coalesce; the necessary meteorological conditions are not common and dense fogs rarely last more than an hour or so in Britain. Combustion products and industrial dust are the common sources of nuclei, so cleaner air means less fog as well as less poisonous "smog". The droplets are very small, 10 to 50 μm (0·4 to 2×10^{-3} in.) in diameter, and may be very numerous, up to 1000 per cubic centimetre.

The meteorologists have classified fogs in terms of the maximum distance (the "meteorological visibility") at which a large black object can just be seen against the sky in daytime, or the distance for which the direct transmittance of a narrow beam of light through the atmosphere is 5 per cent. Surprisingly these two distances are the same within ordinary experimental accuracy. The international classification is shown in Table 16.1.

Whatever the fog density, Koschmeider's law states that the direct transmittance for the visibility distance is constant. However, the transmittance per kilometre varies enormously, being less than 10^{-24} per cent for Code 0, 0·25 per cent to 5 per cent for Code 3 and over 98·5 per cent for Code 9.

The light which is not transmitted directly is, of course, scattered. In a light path equal to the visibility distance, 95 per cent of the light flux will be scattered; by a rather complex extension of this reasoning, if the luminance of a deep daylight fog is taken as

Table 16.1 International classification of fog

Code	Meteorological visibility	Description
0	up to 50 m	Dense fog
1	50–200 m	Thick fog
2	200–500 m	Moderate fog
3	500–1000 m	Light fog
4	1–2 km	Mist
5	2–4 km	Haze
6	4–10 km	Light haze
7	10–20 km	Clear
8	20–50 km	Very clear
9	over 50 km	Exceptionally clear

unity, the luminance of a length equal to the visibility distance will be 0·95. The scattering is not, however, uniform. If a beam of light passes through a small volume of foggy atmosphere, this small volume will become a source of primary scattered light with its own intensity distribution, independent of the distribution in the incident beam. The scattered light will be greatest in the forward direction, least at 90° to 120° from the forward direction and rising towards the backward direction; the relative values of forward, sideways and back scatter are approximately 100:2:5. In practice, of course, there is a lot of secondary scatter so that the sideways and backwards luminance of an elementary spherical volume of fog are much the same at about 15 per cent of the forward luminance.

In daylight the fog is uniformly lit from the upper hemisphere and is of almost uniform luminance, provided that the fog layer is thick enough. Commonly, the zenith luminance may be two to three times the horizontal luminance, but this ratio depends on the reflectance of the ground and becomes unity over a layer of snow, which explains the uniform luminance of the "white-out" phenomenon.

How much light?

In daylight, the reduction in illuminance due to a fog layer is usually less than 50 per cent. Nobody seems to have measured the loss in illuminance under high-mast lighting at night but the early fears that fog would "extinguish" the high-mounted lights were soon proved to be false. Calculation suggests that, if the fog is just thick enough to prevent one lantern being seen from the foot of the next column, the illuminance in a linear road-lighting installation is reduced to about one-third; in an area-lighting installation it is reduced to about one-half. The presence of smoke or dirt in the fog would of course give smaller figures but, in general, fogs do not seriously darken our surroundings, either by day or by night.

Fogs do, however, prevent a lamp itself from being seen, depending on the transmittance per unit distance and the inverse square law. Allard's law dates back to 1864, but is still true and is worth quoting:

$$E \cdot d^2 = I \cdot t^d$$

where E is the illuminance or point brilliance at distance d from a source of intensity I through a fog of t transmittance per unit distance. The threshold of vision of a lamp in dark surroundings is usually taken as a point brilliance of 2×10^{-7} lux, or $0 \cdot 2$ cd at 1 km in clear air; in brighter surroundings a higher point brilliance is required and, very roughly, the value to be taken for E in the equation is $3 \times B^{0 \cdot 75} \times 10^{-7}$ lux where B is the background or sky luminance in candelas per square metre. This becomes a complex calculation, but it is interesting to know the distances at which a 100 cd lamp can just be seen (Table 16.2).

As an example of applying these figures, note that the side lamps of a car are less visible than the car itself in daytime, but are more visible in the dark. Full headlights are advisable in a daytime fog.

Table 16.2 Visual range in fog

Fog code	Met. visibility (km)	Transmission per km (per cent)	Maximum visual range of 100 cd lamp (km) Night	Day
1	0·2	3×10^{-5}	0·51	0·21
4	2	22·4	2·8	0·63
7	20	86·1	10	0·90

The aviation people have three categories of mist and fog which are important in determining the standard of runway lighting etc. on airfields (Table 16.3).

Table 16.3 Fog categories used in aviation

Category	Met. visibility (m)	Night-time range of 100 cd lamp (m)
I	over 800	over 1500
II	400–800	900–1500
IIIa	200–400	500–900
IIIb	50–200	170–500

In category IIIa, a light to be seen at 500 m at night has to be about $2000 \times$ the intensity necessary in clear weather, which calls for severe dimming in good visibility and which causes considerable glare when seen at a short distance in a night fog. Airfield fogs are often stratified and may be either denser or clearer at 50 m height than at ground level. The slope visual range for an approaching aircraft may therefore be different from, and more important than, the visual range along the level runway.

For marine lights, the standard calculations are based on 85 per cent transmittance per km for lighthouses and 89 per cent per km for ships' lights; both these represent clear weather.

On the road, we drive by rule of thumb and optimism, although we call it experience, and we should drive much slower in fog if we did proper calculations. If an average street lamp is just visible at 100 m (about three spans) ahead, the average vehicle rear lamp is visible at only about 50 m which is less than the safe stopping distance from 50 km/h (30 mile/h) on a slippery wet road. A rear light ten times brighter would be seen at 65 m, which would give a better margin of safety in such a fog and a much better margin of safety in less dense fog.

How much contrast?

The principal effect of fog is to reduce contrast. In daylight, colours fade, blacks go grey, whites become indistinguishable from the fog itself; the delicate shading by which we recognize the solidity of things is masked so that the modelling effect is lost and solid objects appear flat. This is, of course, also due in part to the diffusion of the illumination, so that the modelling is in fact worse. Fog does not make things look fuzzy at the edges, although we often get that impression; it reduces contrast by a veiling luminance rather than by blurring the sharp edges of objects. Artists may not accept this fact and may draw "fog" by a ground-glass effect, but the optics of fog do not include small-angle halo effects for transmitted rays of light.

It follows that the only way to improve visibility in a daytime fog is to increase the contrast or the size of the objects to be seen. Blacks and dark colours can be seen farther away than greys and light colours; whites can be reinforced by lights which can increase their actual luminance to many times the natural luminance of the fog and can enable them to be seen at even greater distance than blacks; colours seem to demand this supplementary illumination if they are to be recognized, taking care always that the supplementary lighting does not illuminate the fog and increase its luminance.

The effect of the size of the object is no less important than the contrast effect. In clear daylight conditions, a black-on-white object subtending one minute of arc can just be seen with good eyesight; this is 1 inch at 100 yards or 1 cm at 35 metres. In mist or fog, a small object disappears more readily than a large one, which is one reason why meteorological visibility is defined by a large object. The figures in Table 16.4 are typical for a light fog (Code 3); the last two columns are calculated from the first two.

Table 16.4 Variation of visual range with size of object in daylight

Size of black object	Visual range (m)	Angular size of object (min)	Apparent contrast (per cent)
10 m	550	63	4
3 m	470	22	6
1 m	380	9	10
30 cm	250	4	22
10 cm	140	2·5	40
3 cm	70	1·5	55

Those who have studied the Contrast Rendering Factor in interior lighting would find many similarities in the calculation of contrast in daylight fog.

After dark, conditions are much more complex because the luminous veil and the luminance of the object are both so much more variable than in daylight. To avoid gross reduction in contrast, the light which illuminates the object must be kept away from the line of sight; for an unscreened lamp, the luminance of the foggy atmosphere is inversely proportional to the distance of the lamp from the observer. For a beam of light, it is very much easier to see across the beam or obliquely through it than directly along it, as is commonly experienced with vehicle headlamps and foglamps. The spot lamp on a car should be on the near side and the fan-beam lamp should be on the off side, directed sharply down. For the same reason, high-mast lighting has some advantages over ordinary street lighting and stadium lighting from four towers at the corners is less sensitive to misty conditions than that from just above the roofs of the side stands. It is too complex for discussion here.

Yellow light in fog ?

Old wives' tales often have a factual basis, but this one is very thin. The setting sun is red, distant mountains or the open sky are blue, infrared photography sees through smoke, haze-cutting yellow filters are used for long-range colour photography; all these show that a fairly clear atmosphere transmits yellowish light with less scatter than bluish light. But a cloudy sky is white and infrared radiation does not penetrate fog; if the meteorological visibility is less than a mile or so, all visible wavelengths are scattered and attenuated equally. The range of a vehicle headlamp is reduced if a yellow glass is fitted.

Why does the idea of a yellow foglamp persist? Probably for the same reason that the French use yellow headlamps, to reduce discomfort glare and, slightly, to reduce the recovery time after dazzle.

To be more precise, Rayleigh's law (which may be stated as "range is proportional to the fourth power of the wavelength") is true for particles less than about $0 \cdot 1$ μm in diameter, which means very clear air. For a thin haze with 1 μm diameter water droplets and more than 5 km visibility, there is slight selectivity (range is approximately proportional to wavelength); for mists and fogs as we know them with droplets greater than 2 μm there is no selective transmission.

Industrial smokes and hazes may have very small particles and show some selectivity, which is why some smokes scatter blue light more than yellow and therefore look blue. Few of us remember the London "pea-souper" with its tarry soot particles which absorbed light strongly and coloured it yellow, also scattering it in the usual foggy way. In mist or fog nowadays, there is no coloration and no direct benefit in coloured foglamps if the visibility is less than one mile.

17 Motorway Fog

Most of our manipulative skills were learned from the games we played as children, to kick a moving ball, to build a model, to steer a scooter or bicycle. In all these our eyesight provided the essential information. In later life, we operate these skills best by visual information, perceived subconsciously and used without deliberate selection; this makes it necessary that the information be presented to us in a familiar way. This same thesis may be applied to driving a car—J. M. Waldram has said that "to drive a car in the rush hour on a wet night in a strange town approaches the limit of human capability", or words to that effect.

To most of us, fog on the roads means frustration and delay rather than danger; on motorways, however, we dread the madness of other drivers. Yet most of us drive faster than we ought in fog on ordinary roads, following a white line or judging the line of approaching traffic; on motorways, most of us eventually succumb to the pressures of the surrounding traffic and again drive faster than we ought. What is "faster than we ought"? It means faster than we can see ahead, faster than we can stop if the utterly unexpected happens, faster than we can safely turn to miss a sudden obstruction in the way ahead.

Fortunately the utterly unexpected is very rare, and we get home safely. We follow the vehicle ahead, relying on him to prove that the road is clear and hoping he will not deceive us or do a crash stop. We do this because it is less mental effort than obeying the discipline of really understanding what it is all about.

How we see when moving

The automatic subconscious calculation of speed and distance that we call "experience" is based on clear weather in daylight. We have learned to apply similar criteria to the night-time vision of headlights, rear lights and warnings such as turn indicators or stop lights. Few of us have gained comparable experience in foggy weather, even though some fog occurs on about ten days each year. The tiny clues by which we judge our position, our direction or our speed are very greatly reduced in fog.

E. S. Calvert described the principles of fog vision in his magnificent IES paper "Visual aids for (aircraft) landing in bad visibility" in 1950; a paper that should be compulsory reading for all road designers because the same principles apply to relaxed driving in clear weather. The "parafoveal streamer theory" sounds very off-putting, but it is quite simple—it means that we get our sense of movement from what we see out of the corner of the eye. In clear weather we can identify the distant point towards which we are moving because it is the one spot in the whole visual field which does not appear to move in our frame of reference, such as the edges of the windscreen. The nearer fixed objects in the field of view appear to open out as we get closer to them—this is the ordinary principle of perspective—and eventually they seem to fly past us as we actually pass them. The ways in which these nearer objects move in our field of vision, relative to the fixed frame of the windscreen, give us the clues we need to judge our direction and our speed, quite subconsciously but with great reliability for an experienced driver; they are in fact much more important than the distant and apparently fixed object towards which we are moving. Try it for yourself—you can steer a car quite easily and can judge your speed quite well if a postcard is stuck on the windscreen (even though it is very dangerous in traffic), but steering direction and speed judgement are very difficult if, for example, there is only a small clear aperture in a shattered toughened windscreen. The same effect occurs to a lesser extent with steamed-up side windows.

Calvert described the apparent movement of these nearer objects as "parafoveal streamers" because their optical images stream across the parafoveal region of the retina of the eye, and this is a very fair description for an aircraft pilot using the points of light around an aerodrome runway to guide him in a night-time landing. The parafovea is what is commonly known as the corner or the tail of the eye.

In misty weather, we cannot see as far as our aiming point and we judge our direction by the symmetrical apparent movement of nearby objects. This may seem surprising, but it is true. Directional objects like a white line or a kerb (parallel to our own direction) are real helps, but they are not essential. We tend to fix our vision straight ahead, where there is nothing to be seen, and to control our direction and speed by what we see at angles more than about 10° away from our line of vision, This is the natural method of the learner driver who looks at the end of the bonnet, which is wrong in clear weather but not so wrong in fog. The apparent movement of nearby objects forms a "streamer pattern" similar to that formed by lights at night. We have learned through years of experience to interpret the symmetry or asymmetry and the velocity or changing velocity in terms of our own direction or change of direction and our speed or change of speed. It is the outer parts of the field of view that enable us to move confidently.

It follows that a regular, consistent pattern along the sides of the road will help the driver much more than irregularity. Flashing lights and disconnected objects tend to break the streamer pattern, particularly at night.

Driving in fog

The visual effects are self-evident: distant objects are invisible; nearer objects have low contrast and are therefore less conspicuous; shadows disappear and kerbs merge into the road surface; very near objects seem almost unusually clear. Condensation or frost on the windscreen may be unnoticed but makes things worse.

At night our own headlights fill the fog with a veil of light; reflector studs are very short-range; lights of approaching vehicles appear very suddenly; the vehicle ahead seems to be so clearly visible that we wonder why he isn't going faster; the lights on the car behind are nothing more than a nuisance. The streets lights are an invaluable guide because the buildings etc. are at such low contrast that they are of little help; the flashing character of pedestrian crossing beacons gives them a real conspicuity; unfamiliar lights take longer than usual to be recognized; road direction signs are welcome but they are passed quickly and we dare not slow down to read them.

We seem to move in an enclosed but boundary-less environment in which nothing much happens, frustrated by the absence of clear information. This is a recognized psychological condition leading to tension and poor judgement. We realize the risk of running into something, but we are probably over-confident about our own skill. We can do nothing about the risk from behind, which depends on someone else's skill, and so we seek to minimize the rear collision risk by maintaining speed.

In these circumstances, we do best to keep looking straight ahead, into the fog above road surface level, and to avoid looking directly at the kerb, the white line or the street lights. Let the parafovea do its work; it is more sensitive to change or movement than the central fovea of the eye. Avoid looking at lights, even if they seem interesting. Be mentally alert to sense the existence of an obstruction before you really see it—take advantage of the sub-threshold clues to which the parafovea is sensitive. Keep the rear lights of the vehicle ahead well within their visual range. If this does not allow sufficient stopping distance, let him go. Use well-dipped headlamps or special cut-off foglamps (with clean front glasses) even though the fog may seem to be clearer without headlamps; the safety of the extra light more than compensates for the veiling glare. Use full headlamps in daytime, and bright rear lamps at all times.

All sorts of sophisticated devices have been proposed for giving advance warning of fog conditions. Flashing lights, audible warnings, radio messages and overhead signs might work well when drivers are familiar with their messages, but the most convincing warning is the inability to see something that is known to be there. Drivers may not slow down when they see a sign or notice; they tend to wait until they see a hazard, which may of course be too late. The warning should be given in the same "language" as all the other road information. The design of roads which are prone to sudden fogs should include a succession of recognizable "things", bold but not hazardous, which could become part of the sub-conscious estimation of visibility distance.

Motorway fog

Our motorways are deliberately free from strong contrasts and from frequent hazards; the conditions of seeing are therefore sensitive to weakening of contrasts by mist or fog. The driver on a motorway adopts a longer time-scale than on ordinary roads because speeds are differential rather than absolute and because traffic situations develop relatively slowly. This, incidentally, is why motorway driving is less of a strain and why, after a long spell of it, a driver's decision time is also longer.

At 30 m/s (66 mile/h) the stopping distance at 50 per cent braking effort is something over 120 m, perhaps 150 m. The driver becomes accustomed to looking about 100 m ahead and if the daytime visibility is 100 m he is not likely to slow down. Why should he, if he can see his normal distance ahead? How does he learn that visibility is reduced? There are no lamp columns, no stationary vehicles, no buildings, none of the ordinary clues which give unconscious knowledge of speed or distance. So he does not slow down unless the visibility distance is much less than his usual distance of concern, say 70 m, and as long as this remains consistent he continues in this dangerous situation, able to see for less than 100 m but unable to stop in less than 120 m. This seems silly, but most of us have done it and got away with it because the very safety of motorways is such that we did not need to stop.

Either of two abnormal occurrences can change this blissful travel into disaster. Obviously, a blocked road at 70 m ahead means either a collision or a highly dangerous avoidance manoeuvre. Secondly, a denser zone of fog, perhaps with 50 m visibility, will not be noticed until the driver is in it, now seeing only 50 m but still requiring 120 m to stop. The process can escalate!

The visual information provided on the motorway is inadequate to warn the normal, intelligent, relaxed driver of changing visibility conditions. Flashing yellow lights never work successfully because they are extraneous information calling for a disciplined mental effort; they have not become part of the familiar world of driving, steering and braking. Overhead speed-limit signs are becoming better respected but do not really have an effect until the reason for a speed limitation is made clear. The best thing to tell a driver that visibility is restricted is the sudden appearance of a stationary object at say 100 m ahead which he had not seen when it was 120 m ahead, but the motorway has few such objects. Street lighting installations provide the right kind of familiar stationary objects for the judgement of speed and visibility; it is the existence of the lights themselves and their columns that adds to safety in fog, rather than the light which goes on to the road. Unfortunately, motorways often have cut-off lanterns which are not much use in daytime, even if they are switched on; perhaps the dangers of patchy daytime fog could be reduced by semi-cut-off lanterns or by auxiliary "beacon" lights that are switched on whenever there is mist or fog about.

The night-time fog problems of motorways are much the same in total as those of general-purpose roads; the reflector studs are in the right place but there are no kerbs; the glare is less but the air-light (in which a vehicle ahead can sometimes be seen) is also less; the junction signs are unmistakable but they are so far apart that a driver may wonder if he is lost; the street lighting on motorways is usually well suited to fog driving and gives good guidance. One motorway hazard is worse than on other roads—to follow the rear light of the vehicle ahead is more tempting but is also more dangerous. The driver who has latched on to a tail light dare not stop if he loses his pilot; if the fog thickens, he must accelerate in order to get closer and he finds himself driving even faster than before, in worse visibility.

The features of our motorways which are so advantageous in clear weather are not much help in restricted visibility; in a daytime mist, the ordinary motorway furniture is no help at all. The overhead signals are a genuine attempt to provide helpful information and are the best we have; we shall doubtless learn in time to observe them more respectfully, but a motorway fog still calls for severe mental discipline.

For further reading, the report "Fog and road traffic" published in 1972 by the Transport & Road Research Laboratory (reference LR 446) is to be recommended.

18 Stage Lighting

It was in stage lighting that many budding lighting engineers made their first public experiments in the design, and perhaps the operation, of a lighting scheme. Here they learned the hard shadows given by a single lamp or the flat modelling given by a large number of lamps. Here they learned the different effects caused by a batten-mounted lamp, a flood or a spot, in terms of both the spread of its light and the intensity of its beam. Here they learned to observe lighting effects rather than to look at illuminated objects; to visualize each effect as from the audience while actually watching it from the wings. The discoveries and the disciplines of amateur stage lighting must have converted many an amateur electrician into an amateur lighting engineer.

In this chapter we shall consider the lighting of the professional stage rather than that of the school hall, but the principles are the same. The requirements are so different from those of "ordinary" interior lighting, and so much more exciting, that the "ordinary" illuminating engineer can learn much from his stage-lighting colleague.

Stage lighting must do two principal things; it must reveal and perhaps accentuate the drama of expression and movement of the actors on the stage; it must also reveal or perhaps create the structure and atmosphere of the whole setting. These should be achieved without the mechanics of lighting being obtrusive in any way, whether in terms of glare or colour distortion or false shadows or unnatural changes, unless the style of a particular production decrees otherwise. "Ordinary" interior lighting rarely attempts any of these things but some specialized outdoor installations such as road lighting or decorative floodlighting are designed with no less skill than stage lighting.

Further, stage-lighting design calls for imagination, intuition and experience, in that order. Matters such as lamp efficacy, utilization factor or uniformity ratio are best forgotten, because they have no relation to the visual effect. The stage-lighting designer may have read his codes of practice and his design handbooks, but he leaves them on the shelf.

How much light?

The primary objective is to achieve a three-dimensional flow of light whose intensity is appropriate to the size of the auditorium. There are no standards of illuminance—and none are proposed—because judgements of stage lighting, like all other matters of theatre, are personal and subjective. The level of illuminance at any moment in the theatrical production is therefore determined by the "eye committee" of the production team, with the eye of the lighting director having a casting vote, as a compromise between the differences in judgement and between the optimum levels required for the nearest and most distant members of the audience.

A stage is an environment that has to be looked at rather than lived in; the lighting is done for the benefit of the audience rather than for that of the actors. In most other branches of illuminating engineering, the people who use the light are within the lit environment. Moreover, the stage picture is a controlled picture rather than a random one; the director selects the part of the action that he wants the audience to watch at any given point in the time span of the performance and his principal means of doing this is by light. This selection of dramatic emphasis by means of light is the theatre equivalent of the television director's selection by cutting or dissolving from one camera to another.

Seeing the actors and seeing the scene

Good lighting in any situation, as we have noted in previous chapters, should provide a flow of light to enhance the dimensional quality of the objects in the environment. In a theatrical situation, particularly one involving a picture-frame stage, this dimensional requirement becomes even more important. Actors and objects within the proscenium framework, seen from some distance in larger theatres, appear to lose solidity; their spatial relationship one to another becomes uncertain.

Stage lighting has a particular obligation to make individual actors appear solid, to make individual scenic objects appear solid (even when they are actually flat) and to make all these dramatic elements appear to exist in a dimensional environment. Modern techniques have enabled light to overtake and replace much of the dimensional function of paint and make-up, but this sense of depth is only the beginning of lighting the stage.

Lighting makes a real contribution to the atmosphere of the play; this is largely a matter of colour, much of it based on the gradation from the sadness of cold blue to the happiness of warm gold. Both the selection of dramatic emphasis and the atmosphere are fluid movements throughout the time span of the performance, and they tend to affect the audience at either conscious or subconscious levels according to the rate of change of the lighting. By instant cuts or by progressive changes of 15 to 20 seconds duration to cover a change of mood or area, the audience may have their attention diverted to another piece of stage action or may have their emotional state adjusted without consciously realizing what has happened.

Here lies a basic difference between stage lighting and "ordinary" lighting. When a man enters a theatre, he is willing to allow his mind to be tampered with at a subconscious level by all sorts of means, including light; the same man would be more than indignant if his employer adjusted the office lighting subconsciously to vary the emotional needs of the day's business!

Lamps and equipment

To achieve a controlled picture, stage lighting demands a relatively sophisticated control of both beam and intensity. A light beam that is to be looked at, rather than lived in, requires a high degree of control of beam shape, edge quality and stray light. Historically therefore, there has been an accelerating development away from the simpler floodlighting equipment towards increasingly complex spotlighting.

Most lanterns now in use are designed for, and used with, tungsten lamps. ("Lantern" is the usual name for a luminaire developed for stage use.) Tungsten–halogen lamps are being introduced, often by adapting the existing optics of older lanterns to smaller existing lamphousings, but purpose-designed tungsten–halogen equipment takes proper optical advantage of the smaller size of the new lamp. Meanwhile, as it is standard theatre practice to clean the equipment only when changing the lamp, there is some doubt whether the lumen maintenance during increased life will compensate for the gathering dust on the lenses.

Floods: 200–1000 W GLS lamp with a simple reflector in a box to restrict stray light and to hold the colour filter,

Beamlights: 250–1000 W projector lamp with a parabolic reflector to give a near-parallel beam,

Fresnel spots: 500–2000 W projector lamp with fresnel lens

(prismatic ring type) to give a circular soft-edge beam whose divergence can be controlled by focusing,
Profile spots: 500–2000 W lamp in a sophisticated optical system to give a hard-edged beam whose size and shape can be controlled by internal masks or shutters,
Condenser lanterns: Up to 5000 W, for optical effects such as slide projection.

Lanterns are designed for fixing to standard $1\frac{7}{8}$ in. scaffold bars, either horizontal or vertical and taking account of any restrictions on the burning position. The lanterns are usually angled, focused and coloured for each new production, only their intensity being controlled from a central point. This is a time-and-labour intensive operation and the future may see an increasing development in remote operation, already appearing on television stages. Manually operated follow-spots are used extensively with artistic discretion in theatres following the German technical tradition but in England such follow-spots are used with an obtrusive hard edge mainly in light entertainment, as a mark of status.

Controlling the light

Floods are of limited use because the size of the lit area can only be controlled by varying the distance of the floodlight; this is not a viable method of control because the range of possible lighting positions in a theatre is severely limited by scenery design and auditorium architecture. Nevertheless, compact multi-lamped compartmented troughs wired for several alternative colours have formed the basis of much stage lighting in the recent past. These are called battens when hung above the stage, ground rows when placed on the stage and footlights when placed along the front edge. Although these units are still used in certain types of large musical presentation, they are inappropriate to most productions, particularly on smaller stages, because they provide negative selectivity of dramatic emphasis; the sides of the stage picture receive more light than the centre. In current practice, the principal use of floods is to provide a wash of light over large areas of scenery such as skycloths or painted backcloths.

Lighting the actor requires the more accurate control given by a spotlight whose beam can be controlled by focus and shutter adjustment independently of the distance between the lantern and the actor. The choice of the correct lens and wattage options depends on very rough rules of thumb with wide tolerances. Since it is essential to keep the auditorium as dark as possible, profile spots are used almost exclusively for lighting positions in the auditorium, avoiding any scattered light outside the proscenium picture frame. For many on-stage lighting positions it is easier to use fresnel spots because their soft-edge quality makes them easier to position and to focus, although the scatter from the lens can be an embarrassment if it spills on to the scenery adjacent to the lantern. This scatter can be contained to some extent by a "barndoor" shutter clipped to the front of the lantern but a better method may be to use the more precise profile spot with its lens system set out of focus to give a soft-edge beam. This is, however, a long process and theatre lighting time is an expensive commodity in short supply. Lighting labour costs are not due to high rewards for the theatre-lighting men but are the result of the large number of other staff who have to stand by unproductively while the lighting process is carried out.

The one feature of light which cannot be controlled is beam length. The beam can be shaped, bits can be cut out of the sides or out of the middle, it can be coloured, it can be dimmed, the edge can be softened or hardened but the beam cannot be chopped off in mid-air; it must travel on until it hits a solid object. This is a fundamental problem of stage lighting. The light beam may hit the actor at exactly the correct angle for plastic illumination but the light that misses him passes on until it hits the scenery and makes a splodge of light containing an actor shadow. For truly plastic illumination, the actor must be lit from several angles, giving several shadows on the background; the actor shadows have to be designed and controlled no less than the light which causes them.

Ideally, different light sources would be used for the actor and for the scenery but this is difficult in practice. In the compromise, at some small loss to ideal face lighting, contradictory shadows are put on the floor and on less important parts of the scenery. A vertical downlight does this and can be very dramatic in its emphasis, but it fails to light the actor's eyes and it produces a nose-shadow on his mouth, thus robbing him of his most expressive means of projecting a character. Incidentally, the same is true of narrow-beam downlights used in "ordinary" interiors. Lighting the actor from below causes a most obtrusive shadow which looms above him; horizontal lighting from the front produces a man-sized shadow behind him and also flattens his features. The optimum is to light him from 30° to 60° elevation, to give good modelling of his features whilst producing manageable shadows.

Light at these angles is required from both sides, from the front and from behind to make the actor fully three-dimensional. The balance of the side lights can give a directional key to the scene; the front lighting should be just enough to smooth out the picture without flattening it and the back light separates the actor from his background. The uniform flow of light and the gentle emphasis achieved in "ordinary" interiors would be undramatic in a stage situation and would be quite contrary to the selective requirement of a controlled stage light.

During the performance, the pictorial composition is controlled in terms of selection and atmosphere from a centrally situated dimmer system. This may be highly sophisticated and automated but it only controls the intensity of each lantern; the total effect is dependent on imaginative placing, colouring and focusing.

Planning the lighting

When planning the lighting it is necessary to divide the stage into several controllable areas. There may be a grid plan with 9, 12 or 15 symmetrical areas, depending on the optimum spread of the available lanterns. It is clearer thinking, however, to plan these areas to suit the action of the play; study of the script and discussion with the director and designer usually makes it possible to choose areas which correspond with the light changes even though the symmetrical pattern is lost. Also it is often necessary to double-cover most of these areas, providing two lanterns lighting from each direction, one in a cool colour and one warm so that an emotional range can be obtained by mixing.

This may seem to amount to an alarming quantity of equipment. A simple little play may use upwards of 80 spotlights, but a great deal can be achieved with much less equipment because it all depends on production style. If it is to be an immaculate lesson in realism, the full treatment in multi-lantern plasticity will be unavoidable. Many productions abandon such naturalism and adopt a style in which a handful of lanterns create exciting theatre, but such lighting only works when it is created as an integral part of the production. This is always true; lighting must be conceived as part of the total production and is not something to be grafted on, almost as an afterthought.

Colour

As already indicated, colour forms a major ingredient in lighting style. All lanterns have provision for the insertion of a frame containing the non-flammable plastics which have replaced the coloured gelatines of earlier years. The standard British range has some 70 different tints, and many more can be devised by using sandwiches of two or more. This gives the lighting director all the flexibility he needs and has largely replaced the older technique of using the three primaries (red, green and blue) on separately dimmable circuits to change the tint of a basically white or near-white system. Appropriate colours are chosen for each spotlight, either delicate tints or stronger saturated colours to suit the stylistic requirements of the production.

The open stage

The basic principles of lighting other types of stage, such as theatre-in-the-round, are much the same as for the picture-frame proscenium. Since such forms of open stage are not confined within a two-dimensional frame and have more fluid action and simpler scenery, they require less dimensional assistance from the lighting. Whatever the form of the theatre, the aim remains the same—the illumination should be selective, atmospheric and dimensional to suit the style of the particular production.

In conclusion, we go to the theatre to see the play; even a lighting engineer does not go to see the lighting although it is of fundamental importance to his enjoyment. We may wonder how many other visual experiences would benefit from equal care in the lighting design.

19 Plastics

"A piece of plastics" is a quite proper phrase to use because plastics is a noun, the name given to a whole range of materials. "Plastic" is an adjective meaning easily moulded or pliant, and is much more restricted than the properties of plastics materials. We also use "plastics" as an adjective, as in "a plastics moulding", in our common English way of using nouns as adjectives just as we say a glass lens or a metal bracket; note that "a plastic moulding" would be a moulding that is pliable like clay, which is not quite what we want. So let us have no qualms about saying a plastics diffuser or cable or lamp-holder.

Although there are natural plastics such as rubber and resin, the plastics industry is concerned entirely with man-made, wholly synthetic materials. Some of them have a longer history than we realize; Alexander Parkes made Parkesine mouldings from cellulose in 1861 and Xylonite and Celluloid followed quickly. Baekeland was working on synthetic chemicals for many years before he patented Bakelite in 1907 and synthetic rubbers were produced in quantity in Germany in 1915–20. However, plastics production did not become a significant part of the chemical industry until the 1930's, when it expanded enormously and showed that plastics are good materials in their own right, not just substitutes. That lesson has only recently been properly learned.

Names

"Resin" is a general name for the raw materials of plastics although it is more usually applied to thermosetting pastics such as bakelite resin or polyester resin; it has nothing to do with natural resin from pine trees and it is no good for violin bows.

The individual names of plastics materials range from acrylic, which is a genuine chemical substance, to Zytal, which is a trade name for nylon, known to chemists as polyamide. These names ought to be no more confusing than the strange and varied names of flowers or wines but the relative novelty of these polysyllabic names prevents them from being readily assimilated. Many of the names begin with "poly-" because they are chemically described as polymerized this or that, and we have accepted them as types of plastics (such as polystyrene or polycarbonate) with more real meaning than the trade names. Other names such as Bakelite or Perspex are easier to say or remember than their chemical descriptions (phenol formaldehyde or polymethyl methacrylate). For precision, we should use trade names because they relate to particular forms of each material; for general guidance we should use chemical names because they relate to a range of forms of a particular plastics. There is something to be said for initials (a.b.s. or p.t.f.e. or p.v.c.) which, regrettably, seem to be more memorable than the chemical names that they represent. The non-chemical mind may sometimes confuse polyester and polyethylene, or silicate and silicone; there is no simple rule other than strict mental discipline.

Constitution

The basic constitution of plastics is a micro-structure of interlocking organic molecules, forming chains or networks which give strength and resist permanent deformation. This is quite different from the crystalline structure of metals or the vitreous structure of glass. Glass is not a plastics material, though it does become plastic when hot; the properties of glass and of plastics are quite different in nearly every respect except that they may look the same. Each material has its proper use, with very little competition or doubt if the requirements are properly analysed.

Plastics are of two basic types. Thermosoftening plastics (generally called thermoplastics) soften when heated, rather like rubber or butter; thermosetting plastics harden when heated, rather like clay or bread. When over-heated, the former lose their shape and the latter lose their strength. In general, translucent plastics are thermosoftening while most thermosetting plastics are opaque, though there are exceptions to this generalization; one of the features of plastics is that there are exceptions to nearly every general rule. In luminaires, the choice of the proper plastics material is often determined by its temperature-sensitive properties.

The internal molecular structure of plastics develops during manufacture by a proper balance of pressure, temperature and time. This is the "polymerization" process in which the single molecules in the "monomer" material become knitted together to form the complex molecules of great molecular weight in the "polymer" material. Thermosetting plastics are said to be "cured", usually by moulding from loose powder in hot dies and under pressure to ensure the precise shape; the molecular network extends throughout the whole article to produce a hard or tough material. Thermosoftening plastics are said to "polymerize", which changes the original gas or liquid material into solid as the molecules form long complex chains; this process may require heat or pressure or chemical agents but it always takes time.

The reverse processes can happen to some thermosoftening plastics which may "lose their nature". The internal chemical changes may be initiated for example by ultraviolet radiation or excessive temperature, leading to depolymerization and breakdown of the molecular chains. This can occur during the initial processing or during fabrication if it is not properly controlled, although it is more often due to unfair operating conditions—vapour crazing of stressed mouldings is an old fashioned but still real example. This degradation may not be easily detected but it is sometimes a hidden reason for a batch of mouldings being unexpectedly brittle.

The chemistry of plastics is extremely complex. The principal elements are carbon and hydrogen, with which oxygen, chlorine or nitrogen may be combined. The initial compounds are simple: carbon and hydrogen—methane (CH_4), styrene (C_2H_3), ethylene (C_2H_4); add oxygen—formaldehyde (CH_2O), methyl alcohol (CH_4O), methyl acrylate ($C_4H_6O_2$); add chlorine or some other halogen—vinyl chloride (C_2H_3Cl); add nitrogen—urea (CH_4N_2O) and so on. These are the raw materials of plastics, all non-metallic and essentially organic, based on carbon as the linking atom in the molecular structure.

Most industrial plastics are pure chemical compounds, sometimes blended with other compatible compounds to modify their properties and sometimes mixed with other non-reacting components to act as colorants or fillers. Being pure compounds, they resist the oxidizing effect of air or the corroding effect of dilute chemicals, except of course those liquids which can dissolve or combine directly with one of the compounds. The possible combinations are legion but the purity and the accuracy of manufacture of each product have to be precisely controlled.

Properties

From the foregoing, plastics are obviously light in weight, tough in strength, electrically insulating and capable of being burned. The density is low because plastics are made of the lighter elements and are not compactly crystallized; some plastics will float on water and all are less dense than aluminium or glass. The great merit of toughness—that is resistance to the growth of cracks—arises from the intertwining of the molecules which have to be torn apart before the article will break; some fillers may impair this strength if the granules interrupt the network, but other fillers such as glass fibres may add to the strength if they are well bonded by adhesion to the plastics. Although plastics are manmade, most of them have the toughness which we associate with natural organic materials and they may have very great tensile strength if molecular chains are aligned during manufacture by drawing or extruding through fine dies; the toughness may be lost if they are overcured or internally stressed so that they become brittle.

The electrical insulation arises from the absence of free hydrogen ions or free electrons and from the absence of metals which would provide for their mobility. For much the same reasons, the dielectric losses are low. Heat conduction is also low because the structure is non-crystalline, the atoms being tightly bonded in molecules which have few points of close contact with each other. Another effect of the molecular structure is that pronounced absorption bands occur in the infrared region of the radiation spectrum, so that a plastics diffuser in a luminaire will get hot, partly because it absorbs heat strongly and partly because it conducts heat badly.

Effects of heat

All plastics can be consumed by burning, being organic materials. The right sorts are not dangerously flammable and some are self-extinguishing in the sense that if the source of heat is removed, they go out. The common self-extinguishing ones are p.v.c. and polycarbonates and of course the thermosetting plastics. Those that continue to burn are generally less flammable than wood of comparable thickness.

When thermosoftening plastics burn, they may release a flammable vapour. Acrylics burn cleanly, without smoke, but they may melt and drip; polystyrenes produce a dense black smoke but do not drip as much. The smoke may be the greater hazard. Both kinds have fire-resistant forms but the additives may impair other desirable properties such as the softening temperature. There is also a risk of noxious vapours from some compounds, although these rarely occur in dangerous concentrations.

In lighting, all this is sensibly controlled by recent amendments to the Building Regulations, calling for certain levels of fire resistance to the appropriate British Standards and for certain dimensional limitations for ceiling-mounted or recessed luminaires. Translucent p.v.c. has some merits for large panels or luminous ceilings but must be restricted in thickness so that the total calorific yield is limited. With these relatively new materials, we may have insufficient reliable experience to be able to judge their safety without deliberate testing.

At lower temperatures, below the combustion levels, there are two major effects of heat: to distort the thermosoftening plastics and to reduce the resistance to corrosion and particularly to oxidation of all plastics materials. It is not enough to state a limiting or safe temperature because the time factor is often unknown; a component which safely passes a 100 or 1000 hour laboratory test may fail long before the normal life of a luminaire. Creep, cold flow, oxidation, embrittlement and electrical conduction are all accelerated by temperature rise; this is part of the cost that we have to pay for the real advantages of the peculiar structure of all plastics.

Effects of light and ultraviolet radiation

If radiation is not absorbed, it can do no damage; if it is absorbed in a stable substance, it does no damage. So water-white clear plastics or black pigment-filled plastics would seem to have the best chance of stability.

Although light or ultraviolet radiation may have a direct effect in causing the degradation of plastics, its usual effect is in combination with humidity or temperature or both; again the time factor is critical. For example, the failure of some glass-fibre reinforced polyesters when exposed to the weather has been attributed not to the effect of sunshine or rain on the plastics itself but to the breakdown of adhesion between the two materials and the penetration of the water along the surface of the glass fibre. In lighting, when these several factors are near their joint limit, acrylics are liable to become diffusing, styrenes are liable to go yellow and brown, p.v.c. is liable to go dirty and black and brittle and polythene is liable to go "cheesy" and weak. We have gained experience rapidly in recent years and serious trouble is rare.

Electrical properties

The insulating properties of all plastics are excellent when new and clean; in service the condition of the surface may deteriorate but the internal resistivity remains high.

This is closely associated with non-affinity for water; it does not penetrate into the volume of the plastics and if no surface film develops the insulation is maintained. The effect of atmospheric oxidation reduces the insulation but only very slightly, not usually enough to cause any trouble at mains voltages. Exceptions to this are polyamides and cellulosics of which some forms can absorb significant amounts of water.

The well-known effect of "static"—perhaps more widely

known and feared than is justifiable by the facts—is worst on the best insulators and occurs when electric charges are induced by one of several causes, such as the opening of the moulding die with the wrong release agent or the cleaning of an article with a dry duster or the presence of a new synthetic fibre carpet in the same room. The electric charge induced on the plastics surface does not leak away and it attracts the dust stirred up in the air. That is the story; it does not last long because atmospheric effects soon destroy the extreme insulation, but a better way is to wipe the article with a damp cloth and leave it alone; a very weak detergent may be used.

Conclusion

This chapter has mostly described what plastics are and how they may fail. Between these two extremes, there is a wide and growing range of new, purpose-made materials which give very good service. Plastics are no longer substitutes nor inferior; they have properties of convenience and permanence that we are only beginning to understand and to use. Metal may be cheaper and stronger than plastics for many things, even though it may require protection by a coat of paint which itself is a plastics film with pigment filler, but for many luminaire components there is no substitute that can improve on well designed and well chosen plastics.

20 Glass

Glass is entirely a man-made material, quite different in its constitution from plastics or metal or any other familiar substance. Its nearest common relatives are vitreous enamel and porcelain but both these contain crystals—glass is essentially non-crystalline. It has been with us for over 10 000 years, since long before the Bronze Age, and it has been a household article for 2000 years. It is part of our language and part of our life, whether a looking glass, a spyglass, a drinking glass or just a bottle. It has two principal uses, to enclose or protect something and to redirect light. Both of these are illustrated by a pearl lamp bulb which encloses the gas filling, protects the filament and diffuses the light it produces.

Every glass starts during manufacture as a liquid—very hot, quite clear, electrically conducting and also highly corrosive; it can dissolve anything except the noble metals. As it cools it loses all its chemical activity, its fluidity and its conductivity but it retains some of the desirable properties of a liquid such as clarity, sparkle, and uniformity which respond so well to light. While all metals are basically crystalline and all plastics have a network structure of molecules, glass is scientifically classified as "vitreous".

Glass is a very stiff liquid, of such high viscosity that it can be regarded as quite solid. The laws of nature mean that such a substance must be hard, unyielding, brittle and, if scratched, mechanically weak.

Common glass is made by melting together certain quite common materials, typically sand, soda and limestone, in a fireclay container at around 1500°C. They react chemically, dissolve each other and eventually form a clear liquid, free from bubbles of gas and "stones" of undissolved material. When cooled to about 1000°C it is a treacly liquid, still white hot but stiff enough to be pressed, rolled, drawn or blown into shape. As it cools below red heat, say 500°C, it becomes stiff enough to be handled as a solid. Such great changes in temperature would introduce strains in the solid if they were too sudden, so the cooling has to be gradual below about 500°C—this is the "annealing process".

Composition and properties

Glass may contain any of the known elements except the inert gases, in small or large proportion, and its properties depend on its composition. In glass as in the whole world, perhaps surprisingly, the basic constituent is oxygen. Nearly all glasses are composed of the oxides of metals and semi-metals, not in molecular arrangement but with the oxygen atoms acting as bonds between the atoms of the other elements. The raw materials in the "batch" loaded into the furnace may be carbonates, nitrates or other compounds, but the carbon etc. burns off during the melting process, leaving the oxides. The balance of the elements is stoichiometric (i.e. in the same proportion as in molecules) so there are no unbalanced electrical fields. The proportion of oxygen determines whether the metal atoms are in the reduced or the oxidized state; for example, the iron atoms which always seem to be present as an impurity may be in the ferrous or the ferric state, giving a greenish or a brownish glass.

In 100 grams of ordinary glass, there may be 46 g of oxygen, 34 g of silicon, 13 g of sodium and 6 g of calcium. These are all light elements, so ordinary glass is fairly light in weight, comparable with aluminium. The oxygen atoms are very firmly bonded to each of the other elements, so glass is chemically inert. The weakest bond is perhaps that with sodium—a glass maker may be tempted to include a high proportion of sodium because it makes the glass cheaper, but this will impair the durability. The process of acid etching is largely the leaching out of the sodium by very strong fluorides; glasses which contain more than 16 or 17 per cent of sodium are liable to absorb water from the atmosphere which destroys the vitreous structure near the surface and makes the glass less reliable both mechanically and electrically. Glass fibres, which have a disproportionately large surface area relative to their mass, are made with very low soda content to ensure durability.

Probably the most durable glass is pure fused quartz or silica (48 per cent silicon and 52 per cent oxygen) which can be melted and shaped in much the same way as any other glass but at much higher temperatures. Its special property is a very low expansion coefficient giving easy annealing and great thermal endurance. It is, of course, the enclosure material for high-pressure mercury discharge and for tungsten-halogen lamps. If a fused quartz lamp is finger-marked and then operated, the salt in the finger print will diffuse and react with the silica at the high operating temperature, just as it does in a glass furnace, forming a localized area of higher thermal expansion. If this process goes far enough the local area will contract excessively when the lamp cools and will start a crack.

Other elements may be added to glass to give special properties. A few per cent of boron has the effect of reducing the thermal expansion coefficient without requiring high melting temperatures; a borosilicate glass is a "heat-resisting" glass, of which "Pyrex" is an example. A tiny fraction of iron in the ferrous state—that is without excess oxygen in the batch—imparts a greenish colour and greatly reduces the transmission of near-infrared wavelengths; such a glass is "heat-absorbing" and, if well designed, will have a lower total transmittance for solar heat than for daylight. A tiny fraction of copper may have either of two quite different effects; with an excess of oxygen it gives the usual turquoise colour of copper salts, but a deficiency of oxygen causes the colour to be red, stronger than the red colour of the metal itself. In the old railway semaphore signals, both the red and the green glasses had only copper as the colouring agent. Lead can be used in large proportions, increasing the density and the refractive index to add brilliance and sparkle; lead crystal glass contains about 30 per cent of metallic lead but is water-white. It is worth noting that, unlike plastics, glass contains no organic compounds and relatively little carbon, hydrogen or nitrogen in any form; it is therefore completely nonflammable.

Glass is an excellent electrical insulator when not hot. Such slight conduction as does occur may be due at low temperatures to water adsorption in the surface or at high temperatures to the migration of sodium ions under a strong potential gradient; suitable choice of composition can control either of these effects.

Mechanical strength

Glass has all the properties of a "solid liquid"—extreme strength in compression, limited strength in tension and in shear, and brittleness. With few exceptions, every breakage of a glass article is due to excessive tensile stresses somewhere on the surface of the article rather than inside. Being very hard and rigid, severe stresses can develop at fixing holes or in contact with other hard substances. Any glass component which has to carry more than its own weight should have a cushioned support, with soft washers or cement to prevent localized stresses at glass-to-metal contact.

Brittleness is an interesting phenomenon, not yet fully understood. It occurs in a substance that has microscopic surface flaws, is capable of producing forked cracks, has no internal inhomogeneities to interrupt the growth of a crack and is so hard that plastic flow does not occur. This might be a description of glass.

The tensile strength of glass is greater than one would think, unless one remembers the strength of the $\frac{1}{2}$ mm wall of a light bulb or the atmospheric pressure of about five tons on an evacuated television tube. It depends on the surface condition more than anything else; neither borosilicate glass nor fused silica is much stronger than window glass. The theory is that the surface has a random assortment of submicroscopic flaws which develop unavoidably due to exposure to the atmosphere during manufacture; if a glass rod is drawn *in vacuo* and tested without being touched, an ultimate tensile strength up to 1000 tons/in^2. can be found. This is much greater than the inter-crystalline strength of metals and is comparable with the enormous strength of a pure liquid—but it never occurs in service.

An ordinary figure, at which breakage would not be expected 99 times out of 100, is 1 ton/in.^{2} (150 bar or 15 MPa), greater for blown or fine-etched glass and less for pressed, rolled or sandblasted glass. Fine glass fibres, in which transverse surface flaws have been removed during the drawing process, may have a tensile strength of 10 or 20 times this value, from which we have the great strength of fibre-reinforced plastics. But simple uniform tension rarely occurs in practical glass articles and the effects of their shape, of resilient mountings and of localized loading make calculation uncertain. In any event, there is a statistical uncertainty about any breakage of glass, perhaps $\pm$ 15 per cent in the best laboratories and $\pm$ 40 per cent in service, because of the random nature of the surface flaws.

Glass may be cut to shape by making a controlled flaw in its surface, as with a diamond or a tungsten-carbide wheel, and then taking advantage of its brittleness to propagate a crack along the line of the flaw with very little force. Considering its strength, it is very easy to cut.

"Toughening" may increase the strength of an article by a factor of three or so. The usual process is a heat treatment in which the article is brought almost to its softening temperature, well above the annealing temperature, and then cooled rapidly but uniformly. The surfaces set hard while the middle is still soft enough to prevent stresses developing; as it cools and sets hard throughout, the middle layers will contract more than the surface layers so that the final condition is a sandwich of tension in the middle and compression in the surfaces. The article can then be safely strained until the compressive stresses are overcome, before a tensile failure can occur. Considerable amounts of energy are locked into a toughened article—the compressive stresses may be up to 5 tons/in.²—and so it cannot be cut or ground to size after toughening. Other methods of toughening use chemical treatment of the surface, much easier to apply but mechanically less effective.

Optical properties

Light travels slower in glass than in air; the key measure is the "refractive index", n, the ratio of these velocities. It varies from 1·45 for fused silica to 1·7 for lead crystal and to much higher values for some optical glasses; most soda-lime glasses in common use are about 1·52. Blue light is refracted about 2 per cent more than red, which may be important in optics (coloured fringes around an image) but is negligible in lighting.

The reflection of light at a surface depends partly on the refractive index but more importantly on the angle of incidence; representative figures for sheet glass are 8·5 per cent at any angle up to 40° increasing to 16 per cent at 60° and 40 per cent at 75°. Reflection losses may be high if light passes through a sheet of glass (or plastics) at more than 45°.

A ray of light is bent when it passes obliquely through a glass surface, according to Snell's Law ($\sin i = n \sin r$). The effects cancel at the two surfaces of a sheet of glass but the bending becomes important if the surfaces are not parallel. Examples are a prism or figured glass used in bathroom windows or "frosted" glass which has minute surface irregularities produced by etching, grinding or sand-blasting. This redirection of light can be extremely precise, as in a lighthouse lens, or can be randomly diffusing, as in a pearl bulb; this is the science of optics and much of the practice of lighting.

Whitish glasses do not absorb much light. A usual figure is $\frac{1}{2}$ per cent per millimetre thickness or 4 per cent in a piece of $\frac{1}{4}$ in. plate glass. Tinted or coloured glasses of course absorb much more to give the tint, the transmittance ranging from say 1 per cent for cobalt blue to 20 per cent for signal colours, 50 per cent for some tinted window glass or 80 per cent for a faint amber.

Opal glasses are a "froth" of minute bubbles in the glass; porcelains and glass-ceramics are a mixture of crystalline inclusions in a glass matrix. Both these types of material diffuse and reflect light because of the refractions at the multiplicity of internal surfaces but they do not absorb much light—a suspension of gas bubbles in a clear glass does not contain anything to absorb light. Opal diffusers reduce the light emitted from a luminaire because they reflect some of the light back inside where it may be absorbed, not because they absorb light themselves.

Thermal properties

Glass has a surprisingly low coefficient of expansion, about nine parts per million for 1°C, much less than any plastics. Most common glasses expand and contract less than metals (except tungsten and certain nickel steel alloys). Aluminium expands three times as much as glass. There is a whole field of science in matching metal wires with stable glasses so that they can be sealed together without strain.

The thermal expansion of glass is however sufficient to break it if the temperature is seriously non-uniform. If one part of an article is hotter than the rest, its expansion will put the rest of the article into tension. The local heating may be a time effect when temperatures are changing or may be a steady effect. If there is a uniform gradient from the hot parts to the cool, no stresses develop even with quite high temperatures. It is a non-uniform difference that may cause a breakage. A curved glass can yield under stress more than flat glass, or a glass with straight faces, and can stand greater thermal differences.

Fortunately, glass is a moderately good conductor of heat, nearly ten times as good as plastics, so temperature gradients tend to be levelled out. Moreover, glass freely transmits near-ultraviolet, visible and near-infrared radiation (below 3 μm wavelength) and therefore does not absorb much heat from high-temperature sources. Heat-absorbing glasses, containing ferrous iron, absorb near-infrared radiation quite strongly.

As already indicated, heat-resisting glasses are characterized by low expansion coefficients; to a first approximation, the temperature difference that a given shape of glass will withstand is inversely proportional to its expansion coefficient and to the square root of its thickness.

Permanence

Glass is almost indestructible. An article may be shattered but it is so resistant to corrosion that no natural process will ever dissolve or dissipate a good-quality glass, except by pulverizing it along the sea shore. It is also impermeable and unyielding at normal temperatures. The stories of glass tubes taking a permanent bend when leaning for years against a wall or of mediaeval church windows having the glass eroded by sparrows are imagination unsupported by fact. The only way of getting rid of a glass article is to break it up and put it somewhere else. It can of course be remelted to form a new glass article, and this is welcomed by glass makers whenever they can get supplies of clean broken glass (cullet).

Glass has some affinity for water which leads to the collection of atmospheric dirt, but this can always be removed without damage by the appropriate cleaning agent.

Glass is essential to lighting as we know it today. It encloses and protects, whether in windows or around a source of light—oil, gas or electric. It insulates lamp contacts and high temperature wiring. By reflection and refraction, transmission and diffusion, colour and absorption, it controls the light. To quote Neri (1662): " 'Tis the most pliable and fashionable thing in the world and best retains the form given."

Envoi

Light is an element of life,
a stimulus and a source of pleasure.
Daylight is free
and electric light is relatively inexpensive
but both ought to be respected
and carefully harnessed for our benefit.
The discipline of lighting is a deserving subject
for our study and concern.